# Microelectronics
## In The
# Sport Sciences

Charles F. Cicciarella, EdD
U.S. Sports Academy, Mobile, Alabama

Human Kinetics Publishers, Inc.
Champaign, Illinois

Library of Congress Cataloging-in-Publication Data

Cicciarella, Charles F., 1950-
  Microelectronics in the sport sciences.

  Includes index.
  1. Sport sciences—Electronic equipment. 2. Digital
electronics. I. Title.
GV436.C485   1986        613.7'1        86-287
ISBN:  0-87322-056-0


Editor:  Peg Goyette
Production Director:  Ernie Noa
Typesetter:  Angela Snyder
Text Layout:  Denise Mueller and Lezli Harris
Line Drawings:  Rachel Watson
Photographs:  Charles Cicciarella
Cover Design:  Patricia Phillips


ISBN:  0-87322-056-0

Printed in the United States of America

10   9   8   7   6   5   4   3   2   1

Human Kinetics Publishers, Inc.
Box 5076, Champaign, IL 61820

# Contents

# Preface

The study of sport science, embracing such widely varying fields as physiology, biomechanics, motor learning and behavior, psychology, and ergometrics, often involves the use of mechanical and electronic devices. In a purely instructional setting, where the use of performance measuring devices is completely predictable and limited, such equipment is readily available (though sometimes expensive) from manufacturers. When measurement needs are open-ended, however, as is often the case in research settings, apparatus must frequently be custom designed—a process that can be very expensive.

This book is written to provide the exercise physiologist, biomechanist, sport psychologist, and other sport scientists with an understanding of simple digital circuits and their construction. It begins with an illustrated discussion of elementary digital electronics and circuit construction, written for readers who know little about the subject. Subsequent chapters introduce the use of integrated circuits for counting, logic, timing, and analog-to-digital conversion, and they provide examples of specific applications from a sport or human performance setting. Each example is given in detail, with step-by-step instructions that the novice should be able to follow.

This book is intended solely to serve as an illustration of digital electronics as applied to the human performance area, and should be of interest to researchers who may find it desirable to construct simple circuits. It should not be regarded as a complete treatise on the subject of digital electronics or as a complete examination of human performance applications. Indeed, the field of microelectronics is advancing so rapidly that any attempt at complete coverage would be obsolete before it could be printed.

It is hoped that this book will be useful to professionals in the various sport sciences (exercise physiology, biomechanics, motor behavior and learning, sport psychology, etc.) who need a basic understanding of digital electronics or of the interfacing of laboratory devices to microcomputers. It also should be very useful to students, particularly graduate students in those fields.

# Introduction  [ 1 ]

The laboratory study of sport and human performance is now becoming increasingly sophisticated, especially as that study is conducted in the context of athletic performance, motor learning, and physical education. Such practices as the collection of expired air in Douglas bags, the measurement of heart rates by manual pulse, or cinematographical analysis using pencil tracings on graph paper are much too cumbersome, inefficient, or obsolete for use in serious research. Recent advances in technology, particularly the widespread availability of microcomputers, have enormously improved the speed and precision of data collection and analysis. While the acquisition of laboratory equipment is never without cost (even when paid for by someone else) and can be very expensive, it can be obtained at surprisingly little cost.

If laboratory items such as counters, timers, and various sensors are dissected, they will be found to contain very few parts, especially if modern integrated circuitry has been used in their construction. Actually, for many such devices most of the purchase price represents design work, marketing, and a substantial profit margin. If the laboratory worker could perform the necessary design and construction work instead of ordering a device from a catalog, not only would there be considerable savings but the device also could be customized to meet particular research needs. This is not as difficult as it might seem. This book presents instructions for a series of relatively simple projects for constructing electronic circuits with applications in sport science.

## Terminology

This book is intended for persons with little or no formal training in electronics; every attempt has been made to minimize the use of technical terms and jargon. To help with the technical terms that are used and to assist in the reading of more technical literature, the following explanations of terms are provided:

• Anode — The most positive electrode or connection.

1

- ASCII — American Standard Code for Information Interchange. The code used by most computers to represent letters, digits, and symbols.
- BASIC — Beginner's All-purpose Symbolic Instruction Code. A high level computer language designed to be easy to learn and used by nearly all microcomputers.
- Binary — The base two number system. A system of numbers having only two digits: 1 and 0.
- BCD — Binary Coded Decimal. A system of numbers in which binary numbers up to four digits in length are used to represent the decimal digits.
- Bit — The most basic unit of information used in any binary device such as a computer. A bit may have one of two states.
- Buffer — A device used to isolate circuits from each other. Buffers must always be used between any laboratory device and a computer used to monitor that device. Some computers, however, have built-in buffering circuits.
- Byte — A group of eight bits of digital information. One byte may have 256 different states.
- Cathode — The most negative electrode or connection.
- CMOS — Complimentary Metal Oxide Silicon. A family of integrated circuits based on a particular technological process. CMOS devices are easy to work with, inexpensive, and tolerant to novices. Whenever possible, different families of devices should not be mixed in the same circuit.
- Current — The flow of electrons (electricity) through a conducting substance (usually a wire).
- Digital — Varying in discrete units. In digital electronics all parts of a circuit are either on or off; the actual voltage or current level does not matter to the logic of the circuit.
- DIP — Dual In-line Package. A plastic housing with metal pins on the bottom which contains one or more integrated circuits or other devices.
- Gate — An integrated circuit with two or more inputs and only one output which is on or off depending upon the conditions of the inputs.
- Hexadecimal — The base 16-number system. A system of numbers having 16 digits; 0, 1, 2, 3, 4, 5, 6, 7, 8, 9, a, b, c, d, e, and f.
- High — Positive voltage. With TTL circuits this usually means + 5 volts, although the actual voltage at which a line is considered high is lower.
- Integrated Circuit — A combination of electronic elements that are permanently associated with one another. The term is commonly used interchangeably (but incorrectly) with the dual in-line package containing an integrated circuit.
- Interface — The interconnection of one device to another or of one family of logic to another.
- LED — Light-Emitting Diode. A device that emits light when current flows through it in one direction and does not allow current to pass in the other direction.

- Low — Ground potential or zero volts.
- Microprocessor — A kind of integrated circuit that can rapidly execute a wide variety of functions according to external instructions supplied through programming. The central processing unit of a microcomputer.
- LSTTL — Low Power Schotky Transistor—Transistor Logic. Integrated circuits that are LSTTL technology can usually be used in place of TTL circuits but the reverse may not be true. LSTTL circuits use less current that TTL.
- Printed Circuit — A circuit in which wire connections are replaced by copper tracings on a nonconducting surface whose pattern has been created by a photographic (usually) etching process.
- TTL — Transistor-Transistor Logic. A family of integrated circuits based on a particular technological process. TTL devices are very inexpensive but have some disadvantages, such as heavy current drain, which make them less than ideal when other technologies are available. Whenever possible, different families of devices should not be mixed in the same circuit.

## Some Components

### Dual In-Line Package

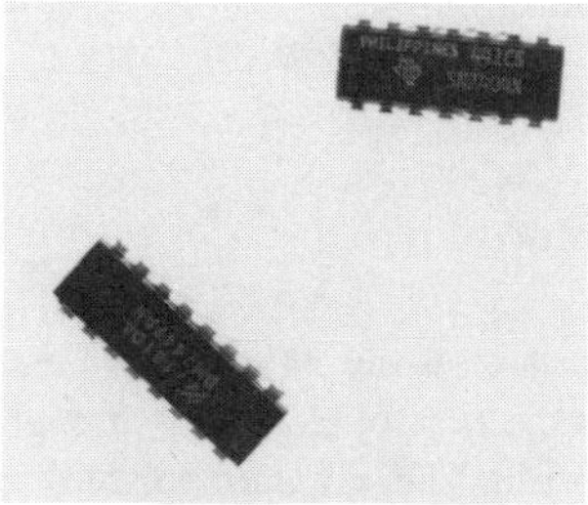

**Figure 1.1** A wide variety of integrated circuits, light-emitting diodes, tiny switches, and other devices are supplied as dual in-line packages which look like plastic centipedes with anywhere from 8 to 40 metal legs.

### Breadboard

**Figure 1.2** A breadboard resembles a miniature sheet of pegboard and is used as a support for all sorts of electronic components. Once a circuit has been designed and tested, the parts may be permanently mounted and soldered on a breadboard or perfboard. Parts are not generally attached to the breadboard itself but are held in place by their interconnections which weave through the breadboard holes.

### Printed Circuit Board

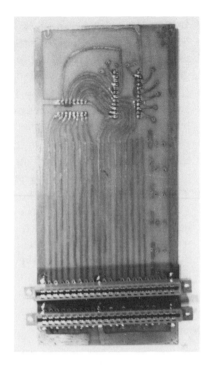

**Figure 1.3**  A complicated circuit may be made much neater, and probably more reliable, by making some of the connections using etched tracings on a printed circuit board. The circuit is "printed" because its image is usually created photographically from a negative, just like a photographic print. The printed circuit, or "PC" board, starts as a sheet of nonconductive material with a layer of copper bonded to one or both sides. The copper is then selectively etched away to leave the desired connections to which various components are then soldered.

### Resistor

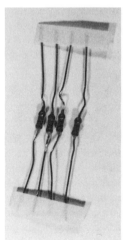

**Figure 1.4**  Resistors are used to reduce or adjust the flow of electricity through parts of a circuit. The resistance provided by each resistor is measured in ohms and is indicated on the side of the device by a code consisting of colored bands. In many cases, resistors with values differing slightly from those shown in a circuit diagram may be substituted with no ill effect. In addition to resistance values, resistors also have different current capacities and will overheat or be destroyed if their capacities are exceeded. All the devices shown in this book use 1/4 watt resistors unless otherwise indicated.

## Handling Integrated Circuits

When any homemade circuit fails to operate as expected, the problem is nearly *always* an improper connection or some other error in design or construction. Only rarely will the problem be a defective or burned part. However, many microelectronic devices can be damaged fairly easily by improper handling.

## Experiment Socket

**Figure 1.5** Temporary or experimental circuits may be easily constructed without soldering by using an experiment socket. Circuit components are simply inserted into the holes in the socket. Each set of five holes is interconnected electrically.

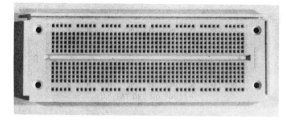

Integrated circuits are especially vulnerable to damage from static. To prevent static damage, always store unused ICs on the conductive foam they are supplied with or on metal sheets. In a circuit all inputs, particularly CMOS ICs, must have a connection to ground, a voltage source, or a signal. When constructing a circuit, never try to solder an IC directly; use an integrated circuit socket instead and insert the ICs only after the rest of the circuit is completed. The pins on an IC and on an IC socket are somewhat delicate and are easily bent or broken. Always make sure that all pins remain straight during insertion and use an IC puller when extracting an IC from a circuit. *Warning*: Trying to remove an IC with your fingers is a sure way to ruin the IC and put about 16 puncture wounds in your finger at the same time.

## Basic Training

The circuits described and illustrated in this book are all quite simple and relatively easy to understand and construct. The real novice may need several careful readings, some experimentation, and perhaps some tutoring. However, this book is not intended to serve as a text or workbook for a basic electronics course unless that course might be specifically focused on sport science applications of electronics. Sport scientists or laboratory assistants expecting to make extensive use of electronics in their work would be well advised to seek a certain amount of formal training in order to take full advantage of the material presented herein. One or two courses in basic electronics, including practical experience in using integrated circuits, would be invaluable. A course in BASIC programming using a microcomputer and a course in assembly language programming (taken after BASIC is mastered) would also prove very useful.

Formal instruction in the above areas is only one means, and possibly not the best, of acquiring skills in electronics. The self-motivated student may find that self-instruction using this and other project oriented books is the most efficient learning path, particularly if highly competent tutors are available for the rough spots.

# Some Basics $\boxed{2}$

Most of this book is dedicated to providing designs and instructions for building a variety of electronic devices useful in the study of human performance. Though these projects can save the human performance researcher a great deal of money, given the cost of commercially manufactured devices, these savings can easily be wasted through spoiled components or unreliable products if poor technique is used in construction. This chapter is intended to provide the novice reader with some basic but essential skills for simple electronics projects.

## Preparing Wire and Components

The proper use of wire can make a major difference in even the simplest project. All wires should be solid, copper wire unless there is a good reason for substitution. For all of the projects in this book, wire from 22 to 26 gauge is acceptable. It is a good idea to be consistent in the size of the wire used and to have on hand a variety of insulation colors.

In preparing wires try to keep them as short as practically possible. This will avoid the development of a spaghetti of wires that makes searching for a wiring error next to impossible; it will also make the finished product look much neater. In some applications shorter wire lengths may even improve the performance of the finished product. Always cut wire with a cutting tool; never break it by bending. A wire stripper makes a good cutting tool and is essential for neatly removing the right amount of insulation from wire ends. To lessen the chance of short circuits, try to remove only a minimal amount of insulation from wire ends; usually 1/4 inch is sufficient.

Components such as resistors, capacitors, light-emitting diodes, potentiometers, and so forth need little preparation before they are used. Lead wires or contact points must of course be clean and grease free before being soldered into place, and they may need to be trimmed or bent so that the component fits securely and neatly into the circuit. To bend a wire or a lead from a component, hold the wire with pliers at the point you want it to bend and then bend the wire

**Figure 2.1** The results of good and bad technique in bending a lead on a resistor.

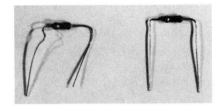

around the curve of the pliers by pressing straight down on the wire. Be careful not to damage the component in any way. Figure 2.1 shows wires that have been bent properly and improperly.

## Using an Experiment Board

Most of the illustrations of circuits in this book show soldered circuits with a breadboard as a base. If you are assembling a circuit simply to learn the procedure, or if a circuit is to serve only temporarily, such a circuit may be easily assembled on an experiment board. This will eliminate the need for soldering and will make the reuse of components more feasible.

The manner in which an experiment board or socket is to be used becomes evident when it is examined carefully. Viewed from the top, the socket has two panels of holes. Each panel has 40 or more rows of holes, each row containing 5 holes. There may be additional sets of holes arranged outside the two main panels. From the bottom of the experiment socket it can be seen (it is best not to remove the paper insulation from the bottom) that each of the five holes in any row is interconnected and that any wire pushed into any hole in a row will be electrically connected to any wires pushed into any of the other holes in that row. Thus it is obvious that the experiment board is used to make circuit connections by pushing wires into appropriate holes in the socket. The interconnections of holes off the two main panels may differ with the particular experiment socket. Frequently all the holes on one side of the socket may be interconnected or they may be connected in sets of five. Such extra holes are usually connected to the power supply or to ground, serving to make such connections readily available wherever needed in the circuit.

The holes in an experiment socket are spaced the same as the pins of an integrated circuit in a dual in-line package. Integrated circuits should be placed directly on the socket (DIP sockets are not needed) so that they span the gap between the two panels of holes.

The connections in an experiment socket are designed to be reused many times provided they are not damaged by misuse. All wires should be between 22 and 26 gauge, free from sharp edges caused by improper cutting, and free from solder or other foreign material. Wires should push in and withdraw easily.

## Using a Breadboard

A breadboard, also called a perfboard, is simply a piece of nonconductive material that serves as a support for the various components of a circuit. The boards used in all the projects in this book have holes of 1.07 mm in diameter and spaced 2.54 mm apart, which matches the spacing on integrated circuit packages.

The breadboard itself serves only as a physical support. Whenever possible, components are mounted on one side and connections (wire wrap or soldered) are made on the other. Components are not bound to the breadboard itself except by the physical structure of the connections on opposite sides of the board.

Integrated circuits are never mounted directly on a breadboard. Integrated circuit sockets are used instead, and the integrated circuits are not inserted into the sockets until the entire circuit is completed. Other components such as resistors, capacitors, potentiometers, and diodes (unless housed in a dual in-line package) are mounted directly on the board.

## Making and Using a Printed Circuit

Building any of the projects shown in this book will probably be most efficient and least expensive if the point-to-point soldering methods described are used. However, on some occasions it may be desirable to use a printed circuit for all or some of the connections in a circuit.

A printed circuit is made by selectively etching a layer of copper or some other conducting material that has been bonded to a nonconducting substrate. Such circuits are highly reliable and have electrical characteristics that can be quite desirable under certain circumstances, but the process of constructing a printed circuit makes them rather expensive if only one or a few copies are needed.

The first step in making a printed circuit is the complete design of the circuit as it is to be etched. This circuit must then be transferred to the unetched surface of the copper-clad board in a manner that will accommodate selective etching. Commercially this is done with photography. A high-contrast photographic negative of the circuit is made and then printed at the proper size onto the surface of the copper-clad board, which has been coated with a substance that becomes insoluble in the etching solution after exposure to bright light. Then the copper-clad board is placed in a solution of ferric chloride which etches the copper where the copper is not protected by exposure to light. After etching, the photo-resist material is cleaned away in another solvent.

For rather simple circuits, the photographic process described above may be eliminated by drawing the circuit (see Figure 2.2) directly onto the surface of the copper-clad board with a special pen that has ink to protect the copper from the etchant, or by using predrawn dry transfers with the same protective properties.

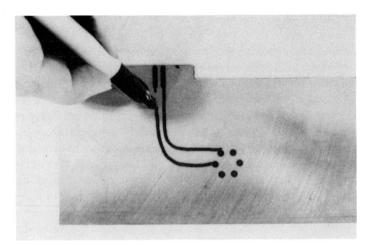

**Figure 2.2** Simple circuits may be drawn by hand directly on a copper-clad surface with a special etch resist pen and rub-on patterns.

The following step-by-step instructions provide more detailed guidelines for etching printed circuits.

1. Draw the resist pattern to scale on paper or a chalkboard unless such a drawing is already available. Mistakes made on the copper-clad board are difficult to correct.

2. Obtain a copper-clad board that is clean, shiny, and free of scratches. Avoid contaminating the surface by handling the board only by the edges.

3. Apply the resist pattern very carefully using rub-on patterns and/or a resist pen. If the photographic method is used, follow the directions for exposing and developing the image supplied by the manufacturer. The edge of a razor blade may be used to gently scrape away resist ink or rub-ons applied in error. Be sure that all rub-on patterns are in firm contact with the copper surface.

4. Place the board *face down* in a tray slightly larger than the board and cover it to a depth of about 1/4 inch with ferric chloride etching solution. Do not be stingy with the etching solution. Too little solution may leave copper where it does not belong, or it may soften the resist material and result in etching of copper in the wrong places. Rock the tray gently to circulate the etching solution without causing the copper board to slide around. The etching process will take about 20 minutes.

5. Discard the etching solution and wash the board with running water for several minutes to remove all traces of ferric chloride.

6. Remove the resist material by soaking the board in solvent (such as isopropyl alcohol) and by rubbing. Allow the printed circuit to dry.

7. Using a drill press with a .033 inch bit, drill the necessary holes. Carefully clean and inspect the completed printed circuit.

## Circuit Construction Hints

Most of the projects in this book may be best constructed using a printed circuit if permanence is desired. The following construction hints should prove helpful to the novice:

1. Components such as integrated circuits, resistors, and capacitors are always mounted on the side of the board opposite the etched copper when a one-sided printed circuit board is used.
2. Always use integrated circuit sockets and insert integrated circuits only after all other construction is completed.
3. Do not trim excess lead wire from components until after the components are soldered in place.
4. Integrated circuits should always be aligned the same way in any circuit.

## Soldering

For constructing the projects shown in this book, you will need a soldering iron with a small, sharp, chisel or pencil type point and a supply of radio type, rosin core solder. Do not use acid core solder or acid paste flux because these may cause damage to electrical circuits.

All surfaces to be soldered must be completely clean and free of grease before soldering is attempted. Also keep the soldering iron tip clean by frequently wiping it with a wet cloth and "tinning" or coating it with liquid solder.

In making a connection, keep in mind that if all surfaces are properly prepared the solder should flow easily toward the heat supply or soldering iron. Place the parts to be soldered in contact with each other and then position the iron so that both parts are heated simultaneously; then touch one or both parts with solder so that the solder flows. Once solder has flowed, withdraw the unmelted solder and the iron and allow the connection to cool. The hardened solder should have a smooth, wet look and the connection should feel and look solid. For difficult connections it may be helpful to tin or precoat each part with solder before placing them into contact. Be careful not to burn any insulation or part of a component (or yourself) by being careless with the soldering iron.

Try to make connections with as little solder as possible, as excess solder looks messy and increases the chance of creating solder bridges or unwanted connections. Excess solder can be removed with a solder braid or suction device.

## Identifying Resistors

Resistors are used to reduce the voltage or current flowing through part of a circuit; the use of the right resistor is very important if a circuit is to function properly and not burn out components. The resistance value of a resistor (in ohms) and the accuracy of that value are indicated by a set of colored bands on the barrel of the resistor. Most resistors have four color bands. The first two (the first three if the resistor has five color bands) represent the first two numbers of the resistance value. The numeric meaning of each color is shown in Appendix C. The next color band is a multiplier for the value represented in decimal by the first two (or three) bands. The last band represents the range within which the value given by the color code is accurate. By using the color code chart in Appendix C, it should be apparent that the color code for a 470 kilohm resistor is yellow, violet, yellow.

## Troubleshooting

When a circuit fails to operate as expected, there are three possible reasons. The first probable cause is an error in the design, that is, the wrong circuit was properly constructed. The second probable cause is an error in construction, that is, the right circuit was improperly wired. Finally, but least likely, one or more components in the circuit is defective or damaged. Unless the components have been mishandled, overheated, or salvaged from old television sets, this is not a likely cause of the trouble. If the circuit design is a new one that has not been tested, and especially if the designer is an amateur, then the design is the most likely culprit and should be checked first. If the design has been previously tested, then the construction should be suspected.

Visually comparing a circuit to a drawing of the intended circuit can be very confusing, especially if the circuit is complex or if construction has been a bit messy. (The first time you have to trace such a circuit you will know why wires should be kept as short and neat as possible.) A systematic approach is needed if this task is to be done with any confidence. One system that might be effective if the circuit is not too elaborate is to check each connection of a given component and then move on to the next component until the error is found.

A logic pen or probe is a device that can be quite helpful in checking the status of each part of a circuit (on or off) under different conditions. It is simply a light-emitting diode with one terminal grounded and the other connected to a probe that can be used to touch any part of a circuit. Whenever the probe touches a part of a circuit that is at a logic "1" (on), the LED will light. A logic source works similarly to a logic probe except that the LED is reversed and connected to a voltage source so that the LED will light whenever the probe tip is grounded.

## Circuit Diagrams

In diagramming an electronic circuit, it is usually not very practical to attempt to show the exact location of each component and wire, or to draw a realistic

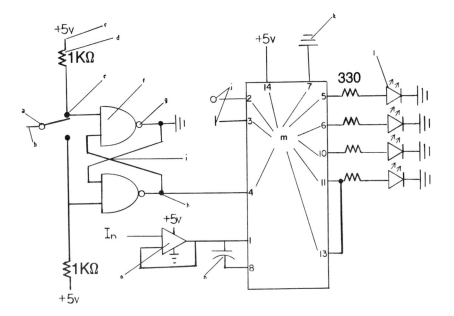

**Figure 2.3** A nonsense circuit illustrating the use of some symbols used in circuit diagrams: (a) a connection that goes low when active; (b) an external connection; (c) a line connected to + 5 volts; (d) a 1000 ohm resistor; (e) one pole of a single-pole double throw switch; (f) a NAND gate; (g) in-dicates the gate goes low when active and distinguishes NAND from AND; (h) shows two lines that are connected; (i) two lines that are not connected; (j) pins tied high (1) and low (0); (k) a ground connection; (l) a light-emitting diode; (m) pin numbers on an integrated circuit.

picture of each component. Instead, the diagram employs certain symbols to represent the various components and to show only the actual connections that are made. A resistor, for example, is shown as a zigzag line with the resistance value printed to one side. An integrated circuit is usually shown either using a symbol for the function performed (such as an AND gate) or as a rectangle with lines coming out representing the required pin connections. One must understand that the connections shown will usually be arranged differently than they are on the actual integrated circuit. Figure 2.3 shows many of the conventions used in circuit diagrams. Some of the symbols used in circuit diagrams to represent various electronic components are illustrated in Figure 2.4.

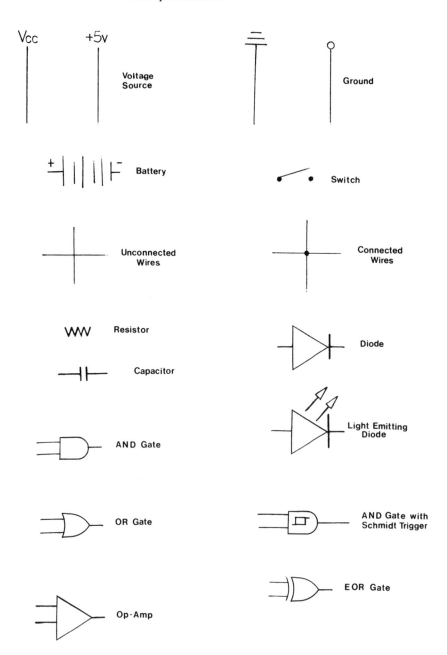

**Figure 2.4** Note that the logic symbols shown above are often used interchangeably with rectangular symbols representing specific integrated circuits, and such rectangular symbols used in circuit diagrams may or may not include the logic symbol. Also, the other symbols shown above are often not the only symbols used for certain devices. Experience and practice is really the only way to learn how to read circuit diagrams.

# Display $\boxed{3}$

The electronic circuits presented in this book generally function perfectly as free-standing circuits. However, this is often not very practical if there is no way their output (the elapsed time, count, frequency, or other data) can be read by a human being. To be useful, the output of a circuit must be displayed so a human being can read it. Electronically coded information may be made readable with displays or devices that transform the electronic information into readable information in the form of visable digits, sounds, or other signals. A wide variety of visual displays such as light-emitting diodes, liquid crystal displays, and computer video displays can easily be used with most of the circuits in this book.

## Light-Emitting Diodes

A diode is a device that allows a current to pass in one direction but not in the opposite direction. A light-emitting diode emits light when a current is flowing through it. In a sense it is similar to a light bulb except that a light bulb is bidirectional. Light-emitting diodes, or "LEDs," are familiar devices used in wristwatches, digital clocks, meters, calculators, programmable microwave ovens, and many other devices. Some advantages of LED displays are that they are low in cost, even in complex displays, and have a relatively rapid response time. Their chief disadvantages are that they create a heavy current drain (ever notice how often your LED watch needs batteries?) and are difficult to read under bright sunlight.

Single LEDs can be round or rectangular or may have other shapes, and they may be equipped with red, white, green, yellow, or transparent lenses. These are often used to indicate the status (on or off) of data lines or other wires in a circuit and can be lit or unlit for either status depending upon how they are connected. Some LEDs even show different colors depending upon the status of the line being monitored. An LED bar graph display is a dual in-line package having eight or more rectangular LEDs mounted side by side.

LEDs do not require much current to operate, and in fact may be damaged if too much current is allowed to flow through them. Also, when an LED is used to monitor the data moving between two parts of a circuit, the LED itself may cause a voltage drop in the data lines if the current is unrestricted, thus altering the data. To prevent damage to the LED, other parts of a circuit, and to the data, resistors must be used to limit the current flow through each LED. The resistance value should be the highest value that still allows the LED to light up sufficiently to be read.

## An Indicator Light

It is often desirable to be able to determine whether a wire in a circuit is carrying a current. With a device for measuring reaction time, for example, we may want to use an indicator light that is activated when the timer is activated as the stimulus to a subject in an experiment. Or, in an unauthorized entry alarm, an indicator light may be used to warn of an intruder or a loss of power to the detector circuit. Figure 3.1 shows a simple circuit diagram for an indicator light using an LED. In this case the LED will be lit whenever there are at least +3 volts in the wire (data line) being monitored. Figure 3.2 shows an alternative circuit diagram for an indicator LED that will be lit when the data line is grounded.

## An 8-Digit Binary LED Display

Circuits that are ultimately meant to be read or controlled by a computer will generally carry data on a group of eight wires called a data bus. Wires that are "on" represent a binary value of 1, and wires that are "off" or grounded represent

**Figure 3.1**  Circuit diagram for an indicator light.

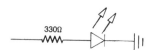
330Ω

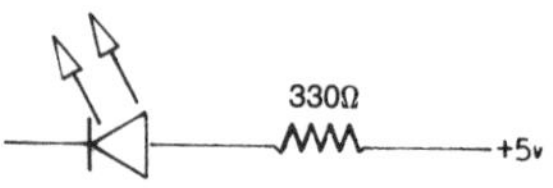

**Figure 3.2**  Note that the LED is a diode and will block current from flowing in the reverse direction. An ordinary light bulb used in place of the LED would allow current to flow in either direction and would emit light with current flowing in either direction. A light bulb would also require much more current to operate.

330Ω
+5v

a binary value of 0. Eight such wires may be on or off in 256 different combinations and can represent binary integers from 00000000 to 11111111. In decimal arithmetic these are the integers from 0 to 255. The output value of such a circuit can therefore be displayed by connecting an LED to each output wire so that the LEDs will be on or off according to the status of the data lines. Each of the eight data lines from the circuit being monitored is connected to a ground line, also from the monitored circuit, through an LED and a current-limiting resistor. The LEDs are connected so that current will flow through them to ground, causing the LED to light up when the data line is in its "high" or "on" state. The "on" state in this case means they carry a potential of at least +3 volts.

The circuit diagram shown in Figure 3.3 illustrates the use of LEDs for reading data from any digital circuit. Each LED is lit when a logic 1 state (+5 volts) is applied to its data input line and is unlit when its input is grounded or unconnected. Figure 3.4 is a photograph of the circuit in which a set of DIP

**Figure 3.3** Circuit diagram showing how LEDs are used for reading data.

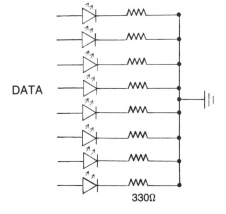

DATA

330Ω

**Figure 3.4** Constructing and experimenting with this circuit should be quite helpful in understanding the use of LEDs. The bar graph display was used to simplify construction, but single LEDs could have been used as well.

switches is used in place of the circuit supplying the data to be read. Physically changing any of the DIP switches simulates the changes that would usually be made electronically. The LED bar graph display used in the photo is simply a set of 10 independent, rectangular LEDs in a 20-pin DIP package.

## LED Decimal Displays

The representation of numbers using the binary system is simple and efficient electronically, but not so simple or efficient when the numbers must be interpreted by people. A number of products contain a set of light-emitting diodes arranged in a pattern such that the decimal digits (or other symbols in some cases) can be displayed by selectively illuminating the proper LED segments. A 7-segment LED display uses seven rectangular LEDs to represent any decimal digit, plus a few other symbols. A 5 × 7 dot matrix LED display uses 35 LEDs arranged in a 5 × 7 rectangular matrix to form almost any character. Multiple digit displays are also available that replicate the 7-segment decimal display in order to display multidigit decimal numbers and perform other functions.

Digital circuits normally represent data in binary. It can easily be seen that the control of seven LED segments to form the 10 decimal digits using such binary input is not a simple task. Fortunately, the task of controlling the seven input lines for a 7-segment display from four binary data lines can be performed by a special integrated circuit known as a BCD to Decimal Decoder/Driver. This circuit has four inputs and seven outputs, the outputs being on or off in the proper combination for controlling a 7-segment display to represent the binary value of the inputs. A decoder/driver circuit that supplies a positive voltage to its output lines to represent the "on" condition is used with a 7-segment common cathode display. A 7-segment common anode display is used with a decoder driver which turns its outputs on by grounding them. Figure 3.5 shows the pinout diagram of two common decoder/driver circuits.

**Figure 3.5**  Pinout diagram of two common decoder/driver circuits.

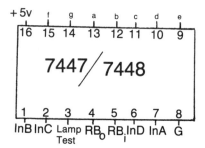

## A 2-Digit Decimal LED Display

Circuits that are meant to yield data to be read by a human being rather than by computer will usually be designed to carry data on a data bus of at least eight wires, but will be represented with the binary coded decimal system. Under this system, each group of four wires will use the binary system to represent digits from 0 to 9. While binary coded decimal is not as efficient in carrying data as the regular binary system, it does allow a decimal display to be generated easily. Each "nibble" of four data lines is connected to the four inputs of a BCD to Decimal Decoder/Driver which is, in turn, connected to a 7-segment decimal LED display. The circuit shown in Figure 3.6 produces a decimal display of the

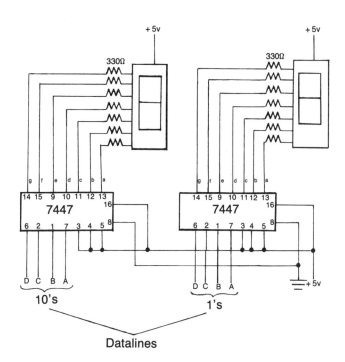

**Figure 3.6** Study the pinout diagrams for the 7447s and the displays. Note that the connections shown in the circuit diagram are arranged for clarity, not for showing actual connection locations. To build this circuit, arrange the displays and drivers as desired on a breadboard or experiment socket and make each pin-to-pin connection as shown in the diagram. Resistors of slightly higher or lower values may be substituted if necessary without major effect on the circuit's operation. Look carefully to make sure all connections are correct and are solidly connected, then test the circuit.

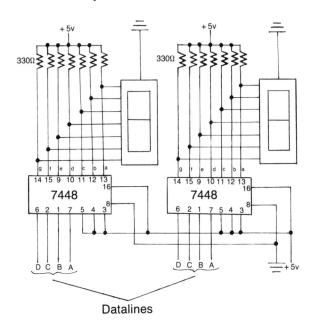

Datalines

**Figure 3.7**  This circuit will be easier to construct if an etched circuit pattern is made for most or all of the connections. This becomes more true as circuits become increasingly complex.

values 1 through 99 using two 7-segment common anode displays, two 7447 decoder/driver circuits, and some resistors. Note that the outputs from the 7447 integrated circuits go high (they are turned on) to light an LED segment and that the resistors are placed between the driver outputs and the display inputs to limit the current flow. It should be obvious that displays of more than two digits can be created by simply replicating the circuit shown here.

Another design for a 2-digit decimal display is shown in Figure 3.7. This circuit uses two common cathode displays and two 7448 decoder drivers, which cause LED segments to light by grounding them. Notice in this circuit that the resistors are used to tie each connection between the driver and the display to a +5 volt source.

## Liquid Crystal Displays

A liquid crystal display is a thin layer of material on a glass substrate that is normally transparent but which becomes reflective or opaque when an electric current is properly applied. Liquid crystal displays, or LCDs, have become quite familiar as the displays in the second generation of digital watches, calculators, and portable computers. LCDs use much less current than LEDs and, because the image they produce is viewed by reflected light, they are easily visible in sunlight. Though their response time is relatively slow compared with other displays and they require a supplemental light source in some situations, LCDs are highly versatile display devices that are widely used in electronic instrumentation.

Because an LCD requires an alternating current source, the circuitry required is a bit more complex than that required for light-emitting diode displays. For the relatively simple devices shown in this book, the use of an LCD should generally be avoided unless the low current drain of the LCD becomes an important consideration. In the LCD circuit which follows, the alternating current source (clock source) is not shown.

LCDs are available in a wide variety of packages from single 7-segment digits to 160 or more character dot matrix modules that are suitable as computer displays. Customized LCDs are also available. LCDs are controlled by integrated circuits called *drivers*, which serve to turn the proper dots or segments on and off and, in the case of multiple digit displays, to make sure the proper digit is displayed.

## A 4-Digit Decimal LCD Display

The FE0202, produced by A.N.D. (a division of Wm. Purdy Co.), is an LCD with four complete half-inch digits plus three decimal points and a colon. The CD4055A is a driver circuit for a 7-segment LCD digital display. (See Figure 3.8 for pinout.)

A 4-digit LCD display may be constructed using the FE0202 and one CD4055A circuit for each of the four available digits in a manner very similar to that used to make the 2-digit LED display. The data to be displayed may come from a timer, a counter, or any other device, but it must be supplied as a binary coded decimal in which ground or zero volts represents a zero value and a value of one is represented by a potential of +5 volts. An alternating current source

**Figure 3.8** Pinout of CD4055A driver circuit.

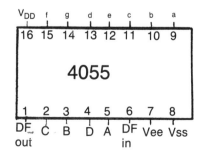

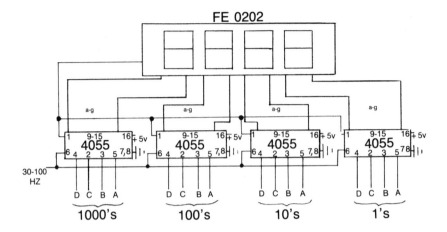

**Figure 3.9** Liquid crystal displays are not as easy to use as LEDs and this circuit should probably not be constructed by beginners. In studying the diagram, note that not all connections are shown; there are actually seven connections between each driver and the display, making 28 connections for the data alone. A printed or etched circuit is essential here.

with a frequency between 30 and 100 HZ (cycles per second) must also be supplied to the display circuit, shown as Figure 3.9.

## Computer Displays

For a circuit whose output is a binary number of eight or fewer digits and in which a 1 is represented by +3 to +5 volts and a zero is represented by ground or 0 volts, it is relatively easy to connect that circuit to a microcomputer so that the computer can read the circuit's output. Once the computer has read a value from such an external device, its own video output can be used in various ways as a display.

Connecting an external or peripheral device to a computer is called interfacing and, while not extremely complex, does require a reasonable understanding of the computer's internal operation if it is to be accomplished properly. Improper interfacing can result in considerable damage to the computer and should not be undertaken by a novice without assistance. For most applications in which the total number of incoming data lines does not exceed eight, a general purpose I/O (input/output) board may suffice and eliminate the need for a custom built board.

For any device sending data to a computer in the form of an 8-bit binary number, and using the proper voltages, an interface circuit must perform two essential functions. First, it must act as a buffer between the computer's data bus and the data lines from the device. This means that the interface circuit must permit the data lines from the device to be connected to the data bus on command from the computer and at *no other time*. If two devices are allowed to enter data on the data bus at the same time, the result will be scrambled data—if in fact the computer continues to function at all. Second, the interface must "latch" the data from the device. This means the interface must momentarily lock or freeze the data lines while they are connected to the computer's data bus so that they do not change while they are being read. Though it may take only a tiny fraction of a second for the computer to read the data sent from the device, this can be a very long time compared to the speed with which data lines can change.

A computer controls where it sends or looks for data through the use of addresses. A typical 8-bit microprocessor such as the 6502 found in the Apple II, VIC 20, AIM-65, or Commodore-64 is capable of addressing 65,536 different locations (2 to the 16th power) through the use of 16 address lines. The binary numbers carried on the 16 lines that comprise the address bus control the location to which the data entered on the data bus will be sent or from where data will be taken. It makes no difference to the microprocessor whether a given location is a byte of RAM or ROM memory, a computer keyboard, a game paddle, or any other peripheral device.

An interface circuit must be designed so that it can detect when it is being addressed in order for it to send or receive data only at that time. This can be done with logic circuitry which uses the 16 address lines as input and supplies the output that sends or receives data only when a specific address is on the address bus. The complete decoding of 16 address lines is not complicated but does require the rather cumbersome use of several integrated logic circuits. Fortunately, most microcomputers have been designed to simplify this process by providing certain control lines that perform part of the decoding; that is, they are turned on or off when certain blocks of addresses are placed on the address bus. The peripheral connectors in an Apple computer, for example, have lines called I/O Select and Device Select which are normally high but are grounded when certain addresses are on the address bus. The memory expansion connector on the VIC-20 has lines called Block 1, 2, 3, and 5 and RAM 1, 2, and 3, which are grounded when certain locations are addressed. While the use of such control lines reduces somewhat the number of devices that can be addressed, such limitations are rarely significant.

Figures 3.10 and 3.11 are examples of simple decoding circuitry for an Apple II and a VIC-20 computer. Because each type of computer may provide a different scheme of control lines, these circuits are not likely to work with other computers. The output of the decoding circuitry should be a line that changes state—from high to low or from low to high—only when the desired address appears on the address bus. This line can then be used to control the buffering circuit. With such circuitry a peripheral device can be read using the command

**Figure 3.10** Though this circuit is quite simple, actual construction is complicated somewhat by the need for etching a *double sided* circuit to match the peripheral connector slot on the Apple II computer. Double-sided copper-clad boards with premasked connector pins are manufactured for this purpose but may be very difficult to find.

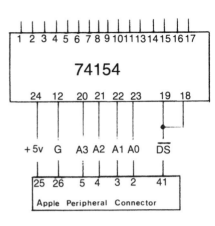

**Figure 3.11** Construction of the VIC-20 interface circuit is complicated by very limited clearance between the interface board and the case of the computer.

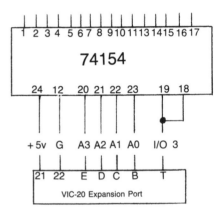

$$X = \text{PEEK (address)}$$

in a BASIC program, or

$$\text{LDA \#address}$$

in a 6502 assembly language program.

The 4508 is a CMOS integrated circuit called a dual quad latch. (See Figure 3.12 for pinout.) It may be used as two separate 4-bit latch circuits or as a single 8-bit latch. Because the 4508 is a 3-state device, it can serve as a buffer as well as a data latch.

Each side of the 4508 has three control pins labeled disable, $\overline{\text{store}}$, and clear. Whenever the disable pin is high (+5 volts) the outputs (labeled Q) are made to float. That is, they are disconnected and will not affect a data bus connected to them. Whenever the disable pin is grounded (made low), the outputs become active. The 4508 is converted from two 4-bit latches to an 8-bit latch by connecting the two disable pins. The $\overline{\text{store}}$ (pronounced "not-store") provides the latching capability. Whenever it is brought low, the outputs will be made to freeze wherever they were when the store was last high. The outputs will follow

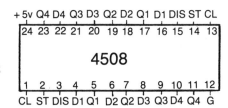

**Figure 3.12** Pinout diagram for the 4508 dual quad latch.

the inputs whenever the store is high. The clear pin is made high to reset all outputs to zero.

## 8-Bit Input Interface for an Apple Computer

A simple 8-bit input interface board for use in an Apple II (trademark of Apple Computer Co.) can be constructed using a single OR logic circuit for address decoding and a 4508 Dual Quad Latch. The inputs to the OR gate are address line 3 and the Device Select control line, and the output is connected to both the store and disable pins of the 4508. The clear pins of the 4508 are permanently connected to ground. With this interface circuit, the eight output lines are made to float except when the microprocessor attempts to read or write data to or from locations $49151 + 16*n$ to $49159 + 16*n$, where n is 8 plus the number of the peripheral connector slot where the interface is connected to the computer. While this is not the most efficient use of the computer's interfacing capacity, and although the circuit does not distinguish between attempts to read or write by the microprocessor, it is a functional and easily constructed circuit. The circuit diagram is shown in Figure 3.13.

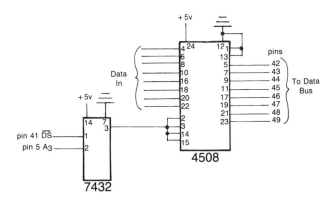

**Figure 3.13** *Turn the computer off* before plugging the interface board into the connector slot. The novice should seek the help of an experienced technician in building this circuit. Incorrect wiring can damage the computer. If the computer fails to boot or does not seem to operate properly, turn it off and inspect the interface circuit again.

Using this interface in peripheral slot #2 of an Apple II computer, the data from a peripheral device can be displayed on the computer's video screen using the following BASIC program.

```
10 HOME
20 PRINT PEEK (49326)
30 GOTO 20
```

## 8-Bit Input Interface for VIC-20

The 8-bit interface described above for the Apple computer may be adapted for the VIC-20 computer with virtually no changes. The only change required is the use of the VIC-20's I/0 3 line in place of the Apple's device select line and a slightly different program for the computer to read the interface. A different style of prototype board must be used, and of course the pin numbers of the data and other lines are different (see Figure 3.14).

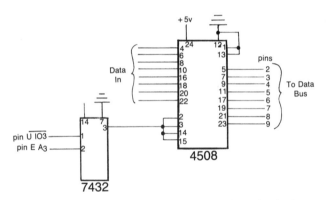

**Figure 3.14** Note all cautions given for the Apple interface circuit above.

## 16-Bit Input Board for Microcomputer

Since most microcomputers have only eight data lines, though many of the newer ones now have 16, an interface circuit for devices producing more than eight bits of data must provide for the transmission of such data in stages. In effect the peripheral device is treated as two peripheral devices of up to eight bits each and are read by the computer sequentially.

Figure 3.15 is the circuit diagram for a 16-bit input board using two 4508 integrated circuits.

Though the address decoding portion of the circuit is for an Apple computer, it should be relatively simple to modify for other computers once the principles involved are understood.

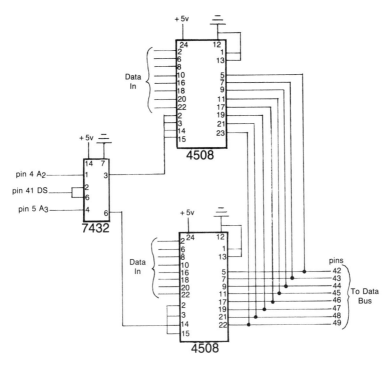

**Figure 3.15** Because of the complexity of this circuit, the use of printed circuit construction for the majority of the connections is recommended. The connections should first be sketched on paper and then transferred to a copper-clad board with resist transfers or a resist pen. The tracings should be arranged so that the jumper wires (need-ed where crossed tracings are unavoidable) are as few as possible. It is important to be sure that all connections to the computer are correct, lest damage ensue to the integrated circuits or even to the computer. If the computer fails to operate with the interface connected, most likely the interface has been improperly wired.

When a 16-bit binary number is transmitted to a computer as two 8-bit numbers, it must be converted back into a single decimal value. The two 8-bit numbers are made from the most significant eight bits and the least significant eight bits of the original 16-bit number. To convert the two 8-bit numbers to a decimal value, the decimal value of the *most* significant 8-bit number is multiplied by 256 and the product is then added to the decimal value of the *least* significant 8-bit number. Thus, if a BASIC program has read the data and stored the most significant eight bits as X and the least significant eight bits as Y, the proper value can be displayed by the command

10 PRINT 256*X + Y

# Counting Devices 4

The function of counting is frequently encountered in nearly every field of research. In the study of human performance the researcher may want to count incidents such as heartbeats, eye movements, revolutions of an ergometer wheel, errors on a motor task, or other events as the dependent measure of an experiment. Counting may also be required as part of an independent variable, such as when a stimuli is to occur whenever some other event has occurred a specified number of times.

Whether driven electronically or otherwise, a counting device must perform three tasks to be useful: detection of the event to be counted, the increment or decrement of a tally, and the display of the current count when required. All of the circuits presented as examples in this chapter show only the detection and tallying functions. Each has an output consisting of a data bus of 4 to 16 lines which may be read by one or more of the circuits shown in chapter 3.

The process of recording a count electronically is accomplished by devices known as "flip-flops," whose output changes state whenever a particular change of state (usually when a positive input changes to ground) occurs. A series of such devices may be put together or cascaded in such a way that each flip-flop provides the input to the next when (and only when) it is itself triggered for the *second* time and returns to its original state, thus producing a device whose several outputs represent a binary number.

A binary counter of any length may be constructed from single flip-flop devices. However, a wide variety of integrated circuits are available in which several flip-flops and a few other devices have been incorporated to form a device specifically designed for counting. Such integrated circuits are easy to use, inexpensive, and widely available, and most have additional features that make them even more versatile. Such features include efficient means of cascading several counters for multidigit counting, resetting to zero, presetting to any desired value, counting up or down, and incrementing by values other than 1. Some counting circuits are even capable of directly driving display devices such as 7-segment LED displays.

## The 7490 and 7493 Integrated Circuit Counters

The 7490 and 7493 integrated circuits are two very popular, inexpensive, and easy-to-use counters. The 7490 is a decade counter (i.e., it has 10 possible output combinations) and the 7493 is a hexadecimal, or "hex," counter (i.e., it has 16 possible output combinations). Both ICs are supplied as 14-pin DIPs (dual in-line packages) and are almost completely compatible with each other. Almost any circuit in which a 7490 is used to count in decimal may be converted to a circuit to count in hexadecimal simply by replacing the 7490 with a 7493. Pinout diagrams are provided in Figure 4.1.

In using either counter the input signal is connected to the Input A pin. The count is incremented every time this source signal makes a transition from high to low and appears at the outputs in binary form. The truth table for each counter appears in Figure 4.2. The output value can be reset to zero (all outputs low) by forcing both reset pins (pins 2 and 3) high. In the 7490 at least one of pins 6 and 7 and one of pins 2 and 3 must be low for the counter to work. In the 7493, pins 6 and 7 are not used.

A particularly useful feature of both the 7490 and the 7493 is that they can be cascaded so that when one counter changes from its highest output value (9 and 15, respectively) to zero, the next counter (next most significant digit of a multidigit) is incremented simultaneously. Any number of counters up to the

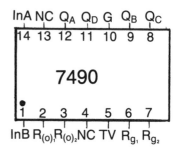

**Figure 4.1** Pinout diagrams for the 7490 and the 7493.

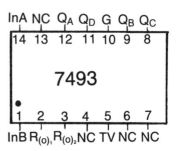

### 7493

| Count | D | C | B | A |
|---|---|---|---|---|
| 0 | L | L | L | L |
| 1 | L | L | L | H |
| 2 | L | L | H | L |
| 3 | L | L | H | H |
| 4 | L | H | L | L |
| 5 | L | H | L | H |
| 6 | L | H | H | L |
| 7 | L | H | H | H |
| 8 | H | L | L | L |
| 9 | H | L | L | H |
| 10 | H | L | H | L |
| 11 | H | L | H | H |
| 12 | H | H | L | L |
| 13 | H | H | L | H |
| 14 | H | H | H | L |
| 15 | H | H | H | H |

### 7490

| Count | D | C | B | A |
|---|---|---|---|---|
| 0 | L | L | L | L |
| 1 | L | L | L | H |
| 2 | L | L | H | L |
| 3 | L | L | H | H |
| 4 | L | H | L | L |
| 5 | L | H | L | H |
| 6 | L | H | H | L |
| 7 | L | H | H | H |
| 8 | H | L | L | L |
| 9 | H | L | L | H |

**Figure 4.2** Truth tables for the 7490 and 7493. H means "high" or a value of about +5 volts. L means "low" or 0 volts or ground.

limits of the available power supply may be cascaded together. Figure 4.3 shows the connections necessary for using the 7490 as a decade or decimal counter.

The connections for the 7493 are the same except that pins 6 and 7 are not used.

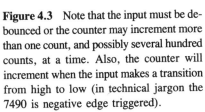

**Figure 4.3** Note that the input must be debounced or the counter may increment more than one count, and possibly several hundred counts, at a time. Also, the counter will increment when the input makes a transition from high to low (in technical jargon the 7490 is negative edge triggered).

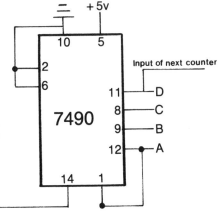

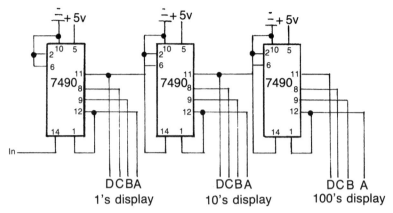

**Figure 4.4** Circuit diagram for a general purpose counter to count in decimals from 0 to 999.

**Figure 4.5** Note that for readability the circuit diagram in Figure 4.4 has the 1s, 10s, and 100s counters arranged in reverse order so that the data lines would have to cross each other to reach the display circuits. The etching pattern also appears to be reversed but is correct because it would be placed on the side of the circuit board opposite the actual display chips.

## A 3-Digit Decimal Incidence Counter

The circuit diagram for a general purpose decimal counter able to count from 0 to 999 is provided in Figure 4.4. Because such a device is likely to be widely used, it is presented again as an etching resist pattern for a printed circuit in Figure 4.5. Connected to a suitable display and an appropriate input signal, such a circuit may be put to a wide variety of applications by the human performance researcher.

## A 4-Digit Hexadecimal Incidence Counter

The circuit diagram and etch resist pattern for a 4-digit hexadecimal (16-bit) counter shown in Figures 4.6 and 4.7 are similar to that of the decimal counter given above.

The 7490s have been replaced with 7493s and the output pins are shown as going to a computer interface rather than to LEDs. Also, the connections to

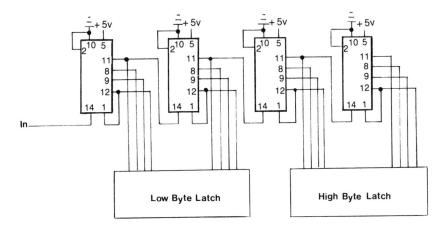

**Figure 4.6** Circuit diagram and etch resist pattern.

**Figure 4.7** Data from the 16-bit counter could be read directly by a computer with a 16-bit data bus. Microcomputers with an 8-bit bus must use software to read the data as two bytes of data and apply appropriate interpretation to the values found.

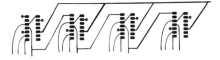

pins 6 and 7 of each counter have been eliminated although they could have been left in without harm. Because hexadecimal rather than decimal is used, the addition of one more counting circuit extends the counting range to 0 to 65535, as compared to the range of 0 to 999 of the 3-digit decimal counter. This is a very significant increase in counting efficiency.

## A Rapid Tapping Board

A rapid tapping board is a device used in the study of motor performance. The subject is given a pen-like stylus with a wire attached and a metal contact plate (or a telegraph key or similar device) and is asked to tap the stylus against the plate as rapidly as possible. The score is the number of contacts (the count) in a specified time period.

Construction of a rapid tapping board is not as simple as connecting a stylus and contact plate to one of the counting circuits described above. An attempt to do so will result in failure for two reasons. First, the counting circuit requires an input that changes between high and low states whereas the stylus and contact plate supply an input changing between high and open or indeterminate states. Second, the counting circuit requires an input that is debounced.

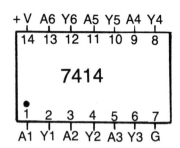

**Figure 4.8** Pinout diagram for the 7414.

**Figure 4.9** Note the small circle next to the output (pin 12). It indicates that the output goes low when the input is made high. This is appropriate for the 7490 and 7493 counters which increment when their inputs make a high to low transition.

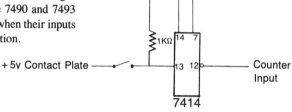

Fortunately, the solution to both problems, though not completely foolproof, is fairly simple. The problem of the inactive state of the input line to the counter (indeterminate when it should be high) is solved by connecting the line to a +5 volt supply line through a 1k ohm resistor. The problem of contact bounce may be solved in most cases through the use of a Schmitt trigger. Schmitt triggers are available in a variety of integrated circuits by themselves and in combination with other logic devices. The 7414 integrated circuit (see Figure 4.8 for pinout) contains six separate Schmitt triggers in a single 14-pin DIP.

Figure 4.9 shows the circuitry required to use the incidence counters already described as a rapid tapping board. The switch shown in the circuit may be a pushbutton, a telegraph key, or a flat, conductive (copper works well) surface and a stylus made from a metal rod.

## Mirror Tracing Apparatus

A mirror tracing apparatus is commonly used in laboratory experiments in motor learning studies. It consists of a flat surface upon which a common shape (usually a star) is drawn in such a way that the area of the shape is nonconductive and all other areas are conductive. This tracing surface is generally placed in a housing where the subject can view it only in a mirror. The subject's task is to trace the pattern with a stylus while straying off the pattern as little as possible.

The circuitry to make a mirror tracing apparatus is exactly the same as for the rapid tapping board already described. The tracing surface may be made by etching away the unwanted copper from a copper-clad board or by selectively covering the copper surface with a pigmented, insulating substance such as nail polish.

## An Alternative Tapping Board

An alternative input counter is a device for counting incidences first at one location, then another, and then the first again. Motor performance researchers may use such a device to test a subject's speed of controlled movement. The subject is given a stylus and the task of alternately touching two contact plates as rapidly as possible for a specified time period.

An alternate tapping board could be made using the simple incidence counter already described for the rapid tapping board and simply adding a second contact plate or telegraph key. However, such a device would be flawed in that two successive contacts at the same contact plate would both be counted. A better solution is to employ some simple logic gating circuits to alternately enable and disable the two contact plates. Figure 4.10 is the pinout diagram for a 7486 integrated circuit which is a quad 2 input Exclusive-OR (EOR) gate.

The output of an EOR gate is high when either of its inputs is high, and low when both inputs are high or low. Figure 4.11 shows how the EOR gate is used to alternately enable and disable two contact plates. One input to each gate is connected to the least significant output bit of the least significant digit of the counter. This bit changes state with each increment in the count. The remaining input to one gate is connected to the +5 volt supply line, and the remaining input to the other gate is connected to ground.

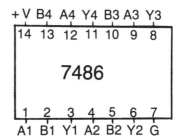

**Figure 4.10** Pinout diagram for the 7486.

**Figure 4.11** In this feedback circuit the output of the counting circuit is fed back to the logic circuit and used to alternately enable and disable two sources of input to the counter circuit.

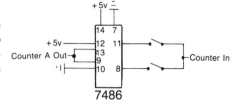

## A Magnetically Activated Revolutions Counter

A magnetically activated switch can be a useful device for counting in any situation in which direct contact between two parts of a mechanical switch is undesirable. The counting of the revolutions of a bicycle ergometer wheel is a good example in that physical contact with a part mounted on the moving wheel might interfere with the movement of the wheel or increase the load on the exercising subject.

A Hall effect switch is a small device that closes a circuit whenever a magnetic field of sufficient strength and correct polarity is encountered. A magnetically activated counter may be constructed using the simple incidence counter presented earlier by simply connecting a Hall effect switch to the +5 volt supply line, ground line, and input line of the counter. Such a counter can become a revolutions counter by mounting a small, permanent magnet on the ergometer wheel (or other revolving device) and attaching the Hall effect sensor where it can sense the magnet with each revolution. Figure 4.12 shows the required connections between the incidence counter and the Hall effect switch.

In high speed or delicate operations where wheel balance is critical, the use of several magnets or some other counterweights may be necessary.

## No-Feedback Steadiness Tester

A steadiness tester is a device for assessing hand steadiness. In two common steadiness testers the subject is asked to hold an electrified stylus between two contact

**Figure 4.12** The output of the switch (marked Counter In) goes high whenever the magnetic field is encountered. As the counting circuits increment on the high to low edge or transition, the counter will not actually increment until the switch is removed from the magnetic field. For the applications described in this book, this is but a technicality that makes little practical difference. The output of the switch could be routed through an inverter if necessary.

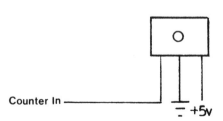

rails a varying distance apart or within holes of decreasing size, without touching the rails or the edges of the holes. The distance between rails or the diameter of the hole between which the stylus can be maintained without making contact is a measure of hand steadiness. Such a device may be made using the incidence counter already described. However, it provides the subject with performance feedback because he or she can feel the contact.

The circuit in the revolutions counter above may be used without modification as a feedback-free steadiness tester. The subject's task is to hold a small, permanent magnet in position over or adjacent to the sensor. One important difference between this steadiness tester and standard devices is that the magnet must be held steady in three planes rather than just two.

## A Light-Activated Counter

For certain very sensitive applications in which even the small force exerted by the magnets of the previous counter is unacceptable or the attachment of a magnet is impractical, a light-activated counter can be a solution. Such a device might be used to count the revolutions of an impeller blade when measuring the volume of expired air or when the close tolerances required by the Hall effect switch are impractical.

For any of the counting circuits already described, a light-activated counter may be made using a photoresistor, a simple device that normally is resistive but becomes a better conductor (i.e., less resistive) when its light-sensitive surface is exposed to light. All that is necessary is to connect the photoresistor to the +5 volt supply rail and the counter input line and then to position a light source and the photoresistor so that the light beam between them will be cut when, and only when, the incident to be counted occurs. This is illustrated in Figure 4.13.

## Some Alternative Light-Activated Counters

For some applications it may be impractical to mount a light source and sensor inside another device to build a light-activated switch for a counter. For example, it is difficult and expensive to build a rotary joint to maintain electrical contact with a rotating device. Such contact may not be necessary, however, if the rotating device can be made to reflect light in a pattern corresponding to its revolution. This might be done by mounting a small mirror inside the device or by

**Figure 4.13** This photo cell circuit is an analog circuit whose output voltage varies with the intensity of the light source. A Schmitt trigger may be needed in circumstances when contact bounce occurs.

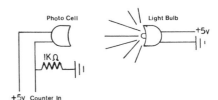

painting different parts of the rotating device black or white. A light beam can then be bounced off the reflecting surface and its changing intensity used to detect each revolution.

## A Pulse Counter

A common task in physiological studies and in fitness laboratories is the monitoring of heart rate or pulse. In the better equipped laboratories this is usually done using sophisticated electrocardiograph (EKG) equipment, which can deliver considerably more information than a simple count. But for some applications such as monitoring workloads of a large group of healthy subjects in an exercise study, the use of such equipment is prohibitively expensive and unnecessary. Measuring heart rate through a 10-second manual count is common and simple in some circumstances, but it can be extremely inaccurate and difficult in other circumstances.

Under those conditions, *photoplethysmography* is far more accurate. It is the measurement of changes in volume of a body part, using light. A light source and sensor are placed on opposite sides of a finger or ear lobe so that light is transmitted through the soft body tissue to the sensor. As the light passes through, the changes in blood volume in the capillaries that occur cyclically with each heartbeat produce small changes in the refraction of the light, and hence in the intensity of the light reaching the sensor. These small changes can be detected with a photoresistor and amplified so they can be used to trigger a counting circuit. Figure 4.14 is a circuit diagram for this. The +5 volt supply line must come from a regulated source such as should be available from a microcomputer. If this circuit is used freestanding, a voltage regulator must be added.

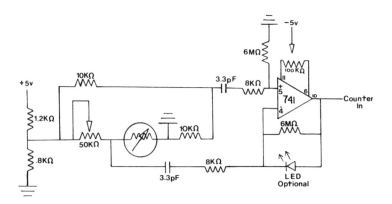

**Figure 4.14** Though this circuit diagram appears more complex than those previously given, this is only a result of having more components. Only one integrated circuit is required, however.

Note that because photoplethysmography is based on fluctuations in the light transmission through soft tissue, the technique is quite sensitive to movement of the body part within the transducing device and is likely to produce faulty data if used in applications other than where such movement is minimized. Though quite useful for heart rate monitoring, other applications require validation with more sophisticated equipment before meaningful data can be collected.

## An Event Sequencer

The alternate tapping board presented earlier is a simple example of a device that can sequence events. The two contact plates are activated alternately with each increment in the contact count. Many experiments may require that three or more such events be sequenced. For example, the alternate tapping board experiment might be expanded to 10 or more contact plates to be tapped in sequence to produce a task involving the sequential use of different muscle groups. Such sequencing might also be used to provide a fixed-interval schedule of reinforcement in studies of learning or conditioning.

In the alternate tapping board circuit, the sequencing was obtained from the least significant bit of the counting circuit since this bit changes with each count. The sequencing of more than two events, however, would require con-

**Figure 4.15** Pinout diagrams for the 8301 and the 8311.

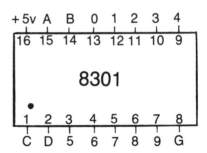

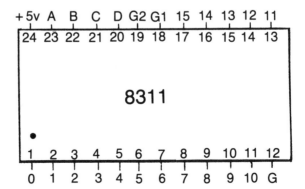

siderable gating circuitry to be controlled by a counter using a 7490 or 7493 IC. Fortunately, there are other integrated circuits that can fully decode the outputs of these counters, and other counters are available with completely decoded outputs.

The 8301 integrated circuit (see Figure 4.15 for pinout) is one of 10 decoders that can accept the BCD output from a decimal counter as input. It has 10 output lines, one representing each of the 10 possible decimal digits. The 8311 integrated circuit (see Figure 4.15 for pinout) performs a similar function but has 16 output lines to decode all 16 possible 4-bit binary values. The circuit design shown in Figure 4.16 shows how the 8301 can be used to sequentially activate 10 different output lines by decoding the output from a 7490 decade counter.

With the exception of the connections to the reset pins of the counter described earlier, the 8301 and 7490 ICs could be replaced by the 8311 and 7493 to produce 16 decoded outputs with no other changes in the circuit. Also, the sequencing of four or eight output lines can be accomplished (see Figure 4.17) by connecting only the A and B or the A,B, and C outputs of the counter to the decoder and connecting the remaining decoder inputs to ground.

The same can also be done by connecting every fourth or eighth output of the 8311 decoder, and the sequencing of five output lines can be done by connecting every fifth output of the 8301 decoder. These connections are illustrated in Figure 4.18.

The sequencing of three outputs is not quite as easily accomplished, however, since neither number of available outputs (10 or 16) is divisible by 3. If it is not necessary to have an absolutely accurate count (i.e., if an error of one count in 16 can be tolerated), it is possible to fool a 7493 counter into providing an output that can be decoded and used to sequence three events. Figure 4.19

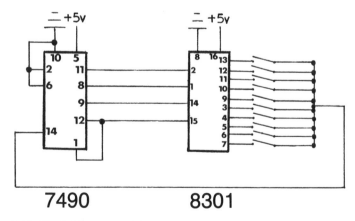

**Figure 4.16** The BCD output of the 7490 is decoded into 10 separate output lines by the 8301.

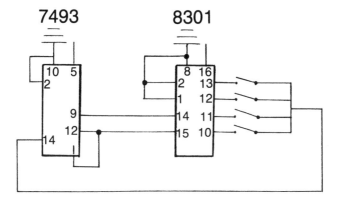

**Figure 4.17a**  The least-significant two outputs of the counter serve as inputs and the two additional inputs are kept low to produce a 4-event sequencer.

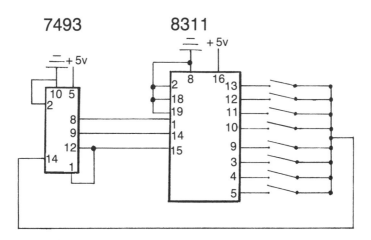

**Figure 4.17b**  An 8-event sequencer using the three least-significant outputs of the counter and an 8311 decoder.

shows how this is done by connecting all four outputs from the counter to a 4-input NAND gate as well as to the decoder inputs.

The output from the gate, whose conditions will be satisfied only once per count cycle, is connected back to the input of the counter so that when all outputs of the counter are high a pulse is generated that will cause the counter to increment to zero. This in effect leaves only 15 output combinations, every third of which can be interconnected to produce 3 output combinations. Technically, the 16th output combination still exists but lasts for only a tiny fraction of a second.

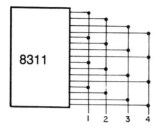

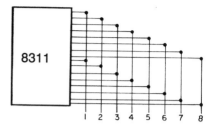

**Figure 4.18** 4-, 8-, and 5-event sequencers can be made from one of 16 or one of 10 decoders by connecting outputs as shown.

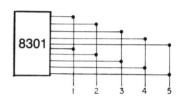

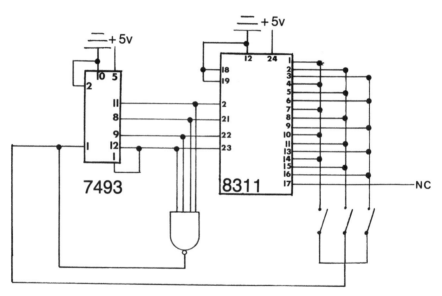

**Figure 4.19** A 3-event sequencer. Every time all outputs are made high, a pulse is generated which resets them low.

## An Alternative Event Sequencer

In situations in which the sequencing of events is more important than the efficient handling of a total count, an integrated circuit counter with fully decoded output may be used. The 4017 integrated circuit (see Figure 4.20 for pinout) is such a counter.

Its output is similar to that of the 8301 but its input is like that of the 7490 or 7493 except that the count increment occurs during a transition from low to high rather than the reverse. Figure 4.21 shows several circuits in which sequences of 10, 5, and 2 events can be controlled using this device.

The 4022 is a similar integrated circuit with 8 decoded outputs and can be used to control sequences of 8, 4, or 2 events.

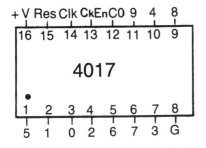

**Figure 4.20** Pinout diagrams for the 4017 and the 4022.

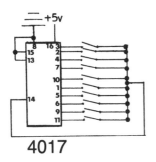

**Figure 4.21** Event sequencers for 10, 5, and 2 events using the 4017.

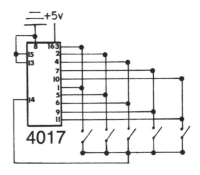

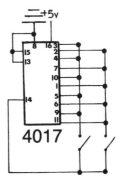

## Countdown Counter With Reset

Experiments in learning or conditioning may occasionally be designed whereby the subject is allowed a fixed number of responses before some form of feedback is given. For example, a subject may be given three chances to find the answer to a problem, with reinforcement given immediately for a correct response and negative reinforcement given after three incorrect responses. Performing such an experiment electronically requires a counter that can be preset to the number of chances to be allowed, is triggered by incorrect responses, and is capable of announcing or decoding when the allotted number of chances have been used. The 74177, 74196, and 4029 integrated circuits are a few counters that can be used to build such a device. (See Figure 4.22 for pinouts.)

The circuit shown by Figure 4.23 can be used to provide a resetable countdown counter using the 4029, which can be programmed for any count up to 16.

The number of permitted trials is entered manually by setting the four toggle switches connected to the load or "jam" inputs. The counter will be set to this value whenever the preset enable pin is made high. The counter is programmed to decrement (count downward) by grounding the up/down pin. The preset enable pin is controlled manually by the researcher or by a separate circuit represented in the diagram by the switch labeled preset. Negative feedback is done through a 4-input NOR gate connected to the counter outputs. The NOR gate is triggered and goes high when all four inputs are low (the count is zero). A 4-input AND gate with a counter that counts upward could also be used in place of the NOR gate. In such a case the preset switches would be set using

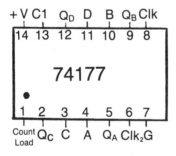

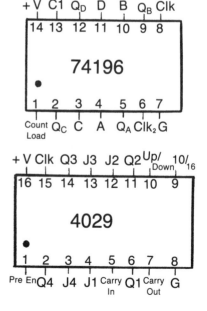

**Figure 4.22** Pinout diagram for the 74177, 74196, and 4029.

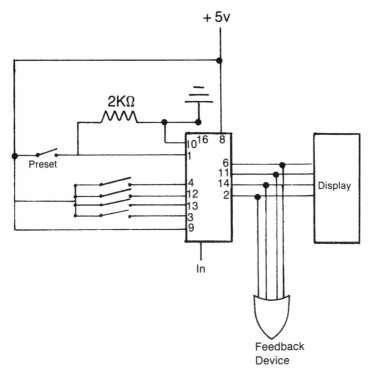

**Figure 4.23** A preprogrammable count-down counter. The toggle switches are used to set the counter to any value up to 16. A feedback device is triggered when the count reaches zero.

negative logic, that is, with the grounded state representing a binary value of 1 rather than zero. With the 4029 this decoding is done on the IC and is available at the carryout pin.

## Counting With a Computer

Many counting applications can be carried out partially or completely by computer. A computer, especially a self-contained personal or microcomputer, can be programmed to receive appropriate input from a variety of devices and to manipulate and/or display a count or some information derived from a count. Usually such programming is simple enough to be done by anyone with minimal knowledge of computer programming.

The use of a computer for counting is easiest when the counting operation can be triggered by a standard peripheral device. Two useful triggering devices that are almost universally available are the keyboard and the game pushbutton. Other triggering devices may include touch-sensing video screens, remote terminals, and various interface boards such as those already presented.

When the triggering device is the computer keyboard (or other remote devices), counting is usually performed using software procedures that form a loop containing a command to increment or decrement a variable. The program is constructed so as to proceed once through the loop each time a specific input occurs. The program generally must also provide some way to display the count after each counting is completed, and some means for ending the counting process. The following example programs are written in BASIC and the purpose of each command is documented with REM statements.

## Counting Program With Running Display

This program counts upward whenever any key is pressed followed by the RETURN key. Pressing RETURN alone ends the program.

```
10  X=0:REM SET COUNT TO ZERO
20  INPUT A$:IF A$=""""THEN 60:REM OBTAIN INPUT, IF INPUT IS NULL
    GOTO LINE 60
30  X=X+1:REM INCREMENT COUNTER
40  PRINT X:REM UPDATE DISPLAY
50  GOTO 20:REM GOTO LINE 20
60  END:REM PROGRAM OVER
```

## Counting Program With End Display Only

This program counts upward every time any key except RETURN is pressed. When RETURN is pressed, the counting ends and the total is displayed.

```
10  X=1:REM SET COUNTER TO ONE
20  GET A$:IF A$=CHR$(13) THEN 50:REM WAIT FOR KEYPRESS, IF KEY IS
    RETURN GOTO LINE 50
30  X=X+1:REM INCREMENT COUNT
40  GOTO 20:REM GOTO LINE 20
50  PRINT X:REM DISPLAY TOTAL
60  END:REM PROGRAM OVER
```

## Countdown Counter

This program counts downward every time a key is pressed until zero is reached.

```
10  FOR X=10 TO 1 STEP −1:REM SET UP LOOP TO COUNT FROM 10 TO
    1 BY ONES WITH EACH PASS
```

```
20  GET A$:REM OBTAIN KEYSTROKE
30  PRINT X:REM PRINT COUNT
40  NEXT X:REM DO NEXT PASS THROUGH LOOP IF ANY REMAIN
50  END:REM PROGRAM OVER
```

With other triggering devices, counting is performed in the same manner but the detection of input may have to be performed differently. Peripheral devices interfaced to the computer through such means as the peripheral connector slots, game I/O connector, or cassette input jack of the Apple II computer or the comparable connections on other computers appear to the microprocessor to be simply memory locations. As such, they are read or written to in the same manner as any memory device. In BASIC, a read from a memory location is performed using a PEEK statement.

An example may prove helpful. The connection for pushbutton number 0 for the Apple II game controller is location 49249. The BASIC command to read the status of the pushbutton is

$$10 \ X=PEEK \ (49249)$$

and will result in the variable X being set to a value of 128 or greater if the button is being pushed when the command is executed, and a value less than 128 if it is not being pushed. The following is a 10-to-1 countdown program written in BASIC and using the Apple II game pushbutton zero location.

```
10  FOR X=10 TO 1 STEP-1:REM SET UP LOOP TO COUNT FROM 10 TO 1 BY
    ONES WITH EACH PASS
20  PRINT X:REM PRINT COUNT
30  Y=PEEK (49249):REM READ PUSHBUTTON
40  IF Y>128 THEN 60:REM IF BUTTON HAS BEEN PRESSED GOTO LINE 60
50  GOTO 30:REM GOTO LINE 30
60  NEXT X:REM DO NEXT PASS THROUGH LOOP IF ANY REMAIN
70  END:REM PROGRAM OVER
```

In assembly language, the BASIC PEEK command is replaced with the LDA command in absolute addressed mode, and the IF...THEN command is replaced by the BIT and BMI or BPL commands.

The memory addresses of the various peripheral connections are virtually certain to differ for every computer and must be found for any given application by referring to the technical documentation supplied by the computer manufacturer. Such documentation should be presented in an understandable format; this should be a major factor in any decision about choosing hardware.

# Timing Devices  5

Timing is probably the function most frequently performed by instrumentation in the study of human performance. Whether the researcher is concerned with reaction times, heart or respiratory rates, frequency of some response, or simply the proper sequencing of events, the fundamental function being performed is often reducible to one of timing.

The passage of time is measured relative to a wide variety of physical phenomena. For example, a year is the time required for the earth to orbit the sun. A day is the time it takes for the earth to revolve about its axis. Devices for measuring almost any predetermined time period may be devised with an understanding of simple physical laws. Knowledge that the acceleration of gravity on earth is 32 feet per second, for example, enables us to measure 1 second by dropping an object from a height of 16 feet (with slight adjustment for air resistance). Understanding of how fine sand flows through a narrow opening makes construction of a sand glass possible. In terms of units of some known measure, the measurement of initially unknown time periods requires that the physical phenomena used as the known measure occur cyclically. Then the measurement of time becomes simply a special case of counting in which the triggering mechanism is the regular cyclic phenomena. Such cumulative timing devices, or clocks, have been built based on pendular motion, balance springs, vibrations of tuning forks, alternating electric current, vibrations of crystals, decay of radioactive isotopes, and even the periods of pulsars.

Electronic timing devices used in sport science are most likely to be based upon the vibrations of a crystal when an electric current is applied, or upon integrated circuits using the charging time of a resistor-capacitor network. Though this sounds rather complex, one does not need a detailed understanding of how such devices work in order to use them with confidence because most of the difficult or complex circuitry is handled in prepackaged integrated circuit form. Integrated circuit products used for timing applications range from monostable multivibrators, or "one shots," to complete stopwatches with display circuits.

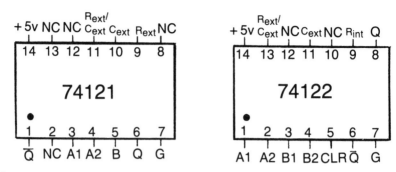

**Figure 5.1**  Pinout diagrams for the 74121 and 74122.

## The 74121 and 74122 One Shot

The integrated circuits designated as 74121 and 74122 are monostable multivibrators (also called one shots). Pinout diagrams are provided in Figure 5.1.

A monostable is a device whose output is stable in one state but not in the other. Whenever the monostable or one shot is triggered, its output changes from the stable to the unstable state and then reverts back to the stable state again. The time period spent in the unstable state is quite precise and depends upon an external resistor-capacitor network; it may range from about 40 microseconds to nearly 30 seconds. Both the 74121 and 74122 have built-in resistors that can be used to charge the external capacitor, or external resistors may be used. The circuit presented in Figure 5.2 may assist in understanding the use of these ICs.

## A Timed Incidence Counter

The rapid tapping board, alternate tapping board, and mirror tracing devices shown in the previous chapter provide only for the counting of incidences (taps or tracing errors), and it is up to the researcher to measure and limit the counting period. These circuits can be made to automatically measure and limit the counting period by the addition of a one-shot circuit producing a pulse for the desired duration and activating the counting circuit. The output of the one shot is used as one input to a 2-input AND gate whose output goes to the contact plate or tracing surface. The line that formerly controlled the contact plate becomes the other input to the AND gate. The circuit diagram is given in Figure 5.3.

## A Heart Rate Monitor

The circuit shown in Figure 5.3 can also be used to measure heart rates or any other parameter that is customarily expressed as a count per minute or other standard time unit. The only requirement is to know the precise data collection inter-

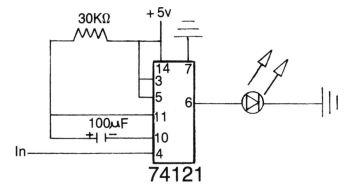

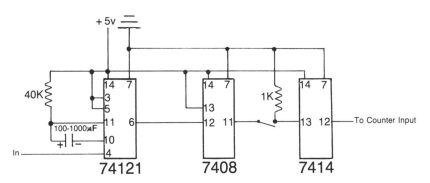

**Figure 5.2** Whenever the input line is made high, the LED will come on for the period determined by the resistor and capacitor values and then return to the off condition. With the values shown for the resistor and capacitor, a pulse width of 2 seconds will result. Other pulse widths may be obtained by changing the resistance and/or capacitance and may be calculated in seconds using the formula

where R is the resistance in ohms and C is the capacitance in Farads. The constant .69 is for the 74121 and is different for the 74122 and other one-shot chips. Exact values must be obtained from data sheets supplied by the manufacturer. Note that the computed and actual pulse widths may not agree exactly, since resistor and capacitor values are subject to some error.

pulse width = .69RC

**Figure 5.3** The switch shown in the circuit diagram represents the tapping board or tracing surface and stylus or other experimental task device.

val provided by the one-shot circuit and the conversion factor for the desired standard time interval. Determining the duration of the one-shot pulse is easily accomplished by connecting the one-shot output to the input of an oscilloscope if one is available. The one-shot pulse can also be measured by using it to control an astable multivibrator or other timing device whose frequency is known. Once the one-shot pulse duration is known, the count per pulse may be converted to a count per standard time unit by multiplying by a conversion factor which is the ratio of the standard time unit to the pulse duration expressed in the same units. The equation for finding the conversion factor for a count per minute is

$$\text{conversion factor} = 60/\text{pulse duration in seconds}.$$

The conversion factor may of course be used in a manual calculation of heart rate or other minute rate, but it will find more frequent and efficient use in a computer program to read the count after each one-shot pulse and display the minute rate. With a one-shot controlled counter interfaced to a computer at location 49326 (a typical interface slot #2 location in the Apple II Computer) and a conversion factor of 10 (6 second one-shot pulse), the following BASIC programs will repeatedly read the counter and display the minute rate. It is assumed that the counter is reset to zero after each reading.

```
10   X=PEEK(49326)
20   X=10*X
30   PRINT X
40   GOTO 10
```
or
```
10   PRINT 10*PEEK(49326)
20   GOTO 10
```

## A Minimum Pace Controller

Certain types of athletic conditioning and many treadmill or bicycle ergometer studies require that the athlete or subject perform at or above some predetermined pace. For example, a swimmer in training may be asked to swim 40 lengths of a pool at the pace of 20 seconds or less per length. If the swimmer moves faster than required for some or all lengths, there is no concern; he or she simply gets a better workout. The only concern is in moving more slowly on any length than prescribed.

The retriggering feature of the 74122 one shot provides a means to obtain feedback any time a predetermined pace is not being met. A retriggerable monostable is one in which the pulse cycle begins anew every time the monostable is triggered. By retriggering before the output pulse is completed, an output pulse of any length may be obtained. By constructing a circuit in which the one shot is retriggered at the conclusion of each performance unit (e.g., pool length, treadmill belt revolution) and using the conclusion of the output pulse to control a feedback device such as a buzzer or light, a low-pace warning device may be constructed. The circuit diagram is given in Figure 5.4

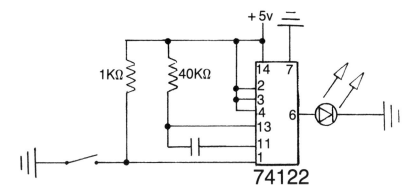

**Figure 5.4** The 74122 has two outputs, one going high for the duration of the pulse and one going low. This circuit uses the inverted output (low pulse) so that the LED will remain unlit during the pulse. Thus the LED comes on to indicate that the performance pace has become too slow and has allowed the pulse to end.

## A Maximum Pace Controller

While it is well known that a certain minimum aerobic workload is needed to produce a beneficial training effect in any individual, too high a workload, particularly as it affects the heart rate, can be dangerous and even lethal for some. For subjects having one or more of the generally recognized coronary risk factors (age over 35, overweight, family history, hypertension, etc.) and wanting to engage in an aerobic conditioning program, it may be helpful to have a device that can detect an overly high heart rate and issue a warning signal. Such a device might also be useful in research studies involving exercise at a prescribed heart rate or involving other rates of occurrence which must be limited.

The minimum pace controller described above can easily be adapted to make a maximum or high pace warning device. In the minimum pace device, the output remains high and prevents the warning signal from occurring as long as a certain triggering rate is maintained; it oscillates, allowing the warning to occur, if the triggering rate is too slow. The actual triggering rate at which the output changes from oscillating to constant may be adjusted by modifying the values of the capacitor and resistors that govern the pulse cycle and can be set to match the maximum desired rate rather than the minimum. The change from an oscillating to a constant output can be detected and used to trigger an alarm by using it as the input to a second retriggerable monostable circuit whose output turns on the alarm whenever it is in the low state.

In use, the first monostable will oscillate in phase with the input signal (heartbeat) as long as the heart rate is below the maximum limit determined by the pulse cycle. That oscillation will keep the output of the second monostable in the high state because the second monostable will be retriggered during its pulse cycle. When the heart rate exceeds the maximum limit, the output of the

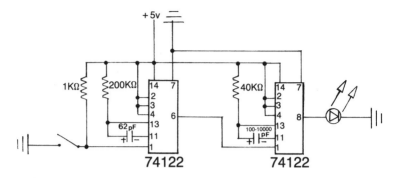

**Figure 5.5** Note that the output of the first 74122 is the inverted output, and the noninverted output is used from the second monostable. Thus the LED comes on to indicate the pace is too fast.

first monostable will be forced to remain high, resulting in the cessation of triggering to the second monostable. The second monostable's output will then go low as soon as its pulse cycle ends, triggering the alarm. The complete circuit is shown in Figure 5.5

## Astable Multivibrators

An astable multivibrator is a device that is unstable in both of two possible states and, consequently, keeps vibrating or oscillating back and forth between these two states. Any device that can be made to do this at a stable rate may be used as the basis of a timer. Figure 5.6 shows a variety of simple astable circuits using commonly available integrated circuits and other simple parts.

The most popular means of building an astable oscillator for timing is use of the integrated circuit known as a 555. Pinouts for this and similar integrated circuits are given in Figure 5.7. By connecting an external capacitor and two resistors, the 555 can be used to build oscillators with a wide range of frequencies. The oscillator circuit is then used as a triggering device to a counting circuit to make a timer.

## A Metronome or Pacing Signal

A metronome or pacing device is useful in many kinds of studies in which a steady pace is required but is not imposed by other equipment. Exercise studies done with a bicycle ergometer or nonmotorized treadmill are prime examples. In such studies the subject may be asked to pedal or walk at the rate of one footfall per beat of a timer. A metronome may be made by controlling a visible or audible signalling device, such as a light, buzzer, or speaker, with an astable oscillator circuit with the desired pulse cycle. In most cases, however, making such a device

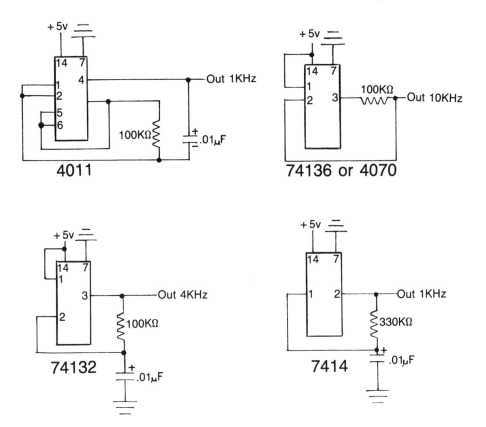

**Figure 5.6** Technical literature available from the manufacturers of these ICs should be consulted for information concerning ac- ceptable combinations of resistor and capacitor values for producing different timing rates.

is not quite as simple as merely connecting the output of the timing circuit to a light or buzzer. If such a connection is made, the light or buzzer will operate in phase with the oscillator circuit, turning on when the oscillator output is in one state and off for the opposite state, as illustrated by the first part of Figure 5.8

It is more desirable for a metronome output to be in the form of spikes of short duration, as shown in the second part of Figure 5.8. One easy way to accomplish this is to use the astable oscillator to trigger a monostable oscillator whose output is connected to the light, buzzer, or speaker. A second desirable feature of a metronome, and one that gives it more versatility, is a varied selection of speed settings. This can be done using variable resistors and capacitors to control the pulse rate of the timing circuit or by using tiny switches to select different fixed resistors or capacitors. A third method is to use decade or hexadecimal counting circuits to divide a fixed pulse rate to produce slower rates. The circuit shown in Figure 5.9 uses switch-selected fixed resistors.

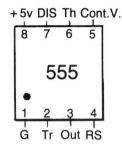

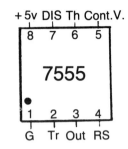

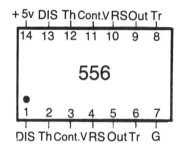

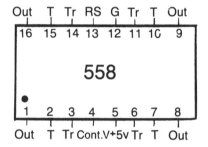

**Figure 5.7** The 555 is a very popular and reliable timing circuit. The 7555 is a CMOS version of the 555 but consumes much less power. The 566 is a pair of timers in a sin-gle IC, and the 558 is four timers in a single IC. For applications requiring only one timer, any of the four devices may be used.

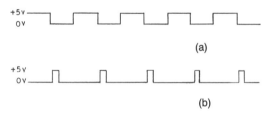

**Figure 5.8** (a) A graphic representation of a square wave generated by an astable os-cillator. (b) Representation of the output needed to make a metronome.

## A Reaction Timer

A very important factor in many sport situations, and a subject of many studies of motor performance, is reaction time or the time required to initiate a volitional response to a stimulus. A device for measuring reaction time must provide an appropriate stimulus and simultaneously start a clock, and then measure the time that elapses before a response begins. In order to separate reaction time from movement time, the response used to stop the timer must involve minimal move-

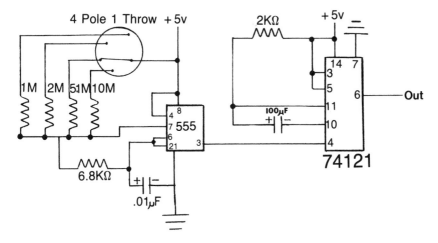

**Figure 5.9** In this improved metronome circuit the output is in the form of spikes of short duration rather than a square wave.

ment by the subject. Usually a light or buzzer is used as the stimulus and a telegraph key or similar switch is used as the response device. The circuit shown in Figure 5.10 shows how a reaction timing device may be built around a 555 integrated circuit and a counting circuit such as one of those presented in chapter 4. This circuit illuminates a light-emitting diode when the timer is started, and a simple telegraph key is used to stop the timer. The switch for initiating the timing process would normally be positioned out of the subject's view.

## A Movement Timer

Another measure frequently taken in the motor performance laboratory, and which may have a bearing upon success in certain kinds of sport activities, is movement time. This is the time required to make a predefined volitional movement such as lifting the hand off a table and swinging the arm 24 inches to the right. Researchers make an important distinction between movement time (the time needed to move) and reaction time (the time needed to initiate movement), which together are often referred to as response time. A clock for measuring movement time must be started in response to the initiation of movement and stopped when the prescribed movement is completed. The basic counter-astable multivibrator circuit already described may be used with the addition of appropriate controls for starting and stopping. For starting the timer, a simple telegraph key that has been electronically debounced will suffice if it is configured to turn the timer on when the key is released. If a controlled movement time is the object of study, a second telegraph key may be used to stop the timer. However, if the parameter of interest is pure movement time, the stop switch must be operable without physi-

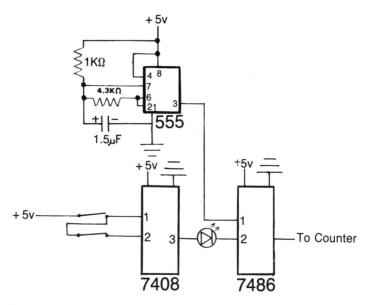

**Figure 5.10** Though the timer in this circuit runs continuously, its output can reach the counter only after switch "a" has been closed and until switch "b" is closed. The resistor and capacitor values shown produce a pulse frequency very close to 100 a second.

cal contact. In the circuit diagram given in Figure 5.11 for a movement timer, the stop switch is operated when a light beam is interrupted.

## A Stabilometer Balance Tester

A stabilometer is a device for testing dynamic balance in a standing subject; it is used frequently as a novel task in motor learning studies and may have some value in predicting success in sports such as skiing and water skiing, in which standing dynamic balance is critical. It is simply an unstable teeter totter platform upon which the subject stands. Most studies require a subject to try to keep the platform balanced and then time how long it is actually in balance during a test period of perhaps 30 seconds. Such a data collection task can be accomplished very accurately using a 555 timer adjusted to a frequency of about 100 Hz (100 cycles per second) to trigger a counting circuit whenever the stabilometer is on or off balance. The circuit to do this is shown in Figure 5.12. A counter such as the 4-digit binary counter described in chapter 4 is triggered by the output of a 2-input NAND gate. The inputs of the NAND gate are the output from the 100 Hz timing circuit and the output from a switch that is closed when the stabilometer is balanced. The timer circuit, then, runs continuously but its signal only reaches the counter when the stabilometer is balanced. A mercury tilt switch is an obvious choice for sensing the position of the stabilometer.

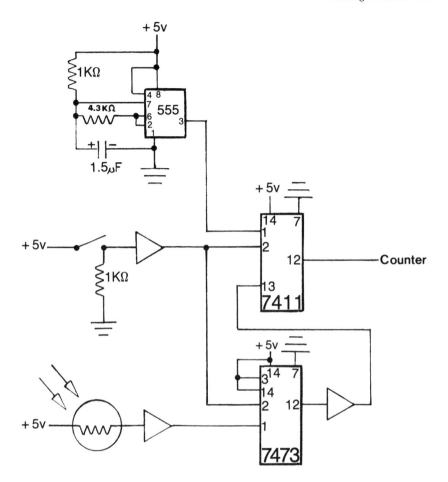

**Figure 5.11** The output of the timer is allowed to reach the counter from the time switch "a" is opened to when switch "b" is closed. For clarity, separate logic symbols have been used in the diagram to indicate the inverters even though all three are available in a single integrated circuit.

## A Pursuit Rotor

A pursuit rotor is a novel task device used in studies of coordination. It consists of a rotating surface similar in design and speed to a phonograph turntable, with a smaller "target" surface located several inches away from the center so that the target moves in a circle. More elaborate devices may also be constructed in which the target moves in an elipse or other pattern. The subject is told to keep a stylus or sensor in contact with the target as it moves for as long as possible. The circuit to measure the time on or off the target electronically is identical to that shown for the stabilometer balance tester above, except for the design of the sensor.

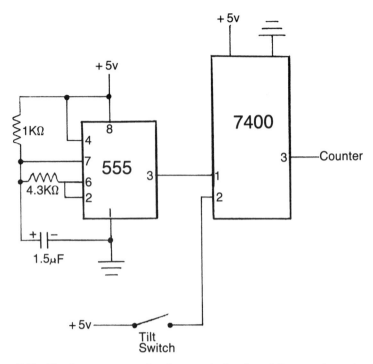

**Figure 5.12** The timer output reaches the counter whenever the tilt switch is closed. Depending upon how the tilt switch is at- tached to the stabilometer, it may be closed when the subject is on or off balance.

Because electrical contact with a rotating device can be difficult to main- tain, it is not practical to wire the target as a contact plate as was done with the tapping board above. A better solution is to use a magnetically activated (Hall effect) switch mounted in the tip of the sensor and to attach a permanent magnet to the target area. However, the researcher doing this must be aware of the fact that the effective size of the target area will not be the area of the magnet but the area over which the magnetic field is strong enough to trip the switch. Another practical solution is to mount a light source and battery on the rotating surface and use a photodetector in the sensor device.

## Timing by Computer

Many studies in which the end of the timing period can be detected by a device connected to a computer (such as a keyboard) can have timing and related func- tions performed under software control. The computer can handle a variety of stimulus and feedback conditions with ease and can be easily programmed to perform many kinds of timing operations as well as saving, retrieving, displaying, printing, and even analyzing the data obtained. This is not to say that the computer should be used in all cases, however. The computer, particularly the personal size computer most likely to be used for studies of this type, does have limitations that can quickly undo data collection in some instances. These limitations may stem from inadequate (for the application at hand) design features of the computer itself such as its graphics, color, sound, or BASIC language, but usually they are more fundamental in nature. Microcomputers and microprocessors employ digital electronics that sequentially execute a variety of instructions supplied by a computer program. The execution rate is precisely controlled by a crystal oscillator with a frequency (often called a clock rate) ranging from approximately .5MHz to 4MHz (1MHZ = 1 million cycles per second).

A computer program can employ a loop to perform timing functions if the number of clock cycles required for execution is known. Unfortunately, because BASIC programs are executed in such a way that the exact execution time cannot be predicted easily, timing programs must be written in the much more difficult assembly language (also erroneously called machine language) if extreme precision is needed. Furthermore, any program branches that may occur during a timing loop all require the same number of clock cycles for execution to maintain extreme accuracy. Another related problem stems from any program steps inside the timing loop which are not directly concerned with the timing process. Common examples of such steps are commands to update the computer's display for providing subject feedback of elapsed time. Such commands can easily consume 90% or more of the execution time for each cycle of the timing loop, during which time the computer cannot respond to changes in its input devices.

## Built-in Computer Clocks

Some microcomputers and most larger computers have built-in peripheral devices for measuring time in the familiar units of hours, minutes, and seconds. These are designed to keep accurate time without interfering with the operation of programs and without placing a heavy demand on microprocessor operating time. Add-on peripheral boards are usually available for the same purpose for computers (such as the Apple II+) which do not have such clocks as standard equipment.

Whether the clock is built in or is added to the basic computer system, it will always require a constant power supply to operate. Thus, the computer system using a clock must be operated 24 hours per day or the clock must be provided with a separate, rechargeable battery to keep it operating when the computer is turned off. Of course, if keeping track only of elapsed time is more important than tracking the actual time of day, such clocks can simply be reset whenever the computer is turned on. The technical manuals supplied with the computer or add-on board should be consulted for details.

## Software Clocks

For computers that are not equipped with built-in clocks, an elapsed time clock function able to keep track of hours, minutes, seconds, and even small fractions of seconds can be implemented with software if an appropriate oscillator or clock pulse signal is available. Many microcomputers are equipped with special integrated circuits called interval timers. Such circuits (some examples are the 6522, 6530, and the 6532) can be easily programmed to time any interval from 1 to 65536 microprocessor clock cycles (approximately 1 to 65536 microseconds) and to send an interrupt signal to the microprocessor after each timing interval. Similar signals can be generated by peripheral boards with computers that lack these special circuits.

When the microprocessor receives an interrupt signal (most can be interrupted in several ways), it will finish executing whatever instruction is currently being processed and then check a special memory location for the starting address of a program (entered by the user) for handling the interrupt. The interrupt handling program can be a software clock program or almost anything else. Provided the frequency of the interrupt signal is not too high (more than about 100 cycles per second), the time demand placed on the microprocessor by such programs will be small and should not produce an objectionable or even noticeable effect on the execution speed of other programs. Although the implementation of an assembly language clock program is not particularly difficult, certain details must be worked out individually for every different microprocessor and computer configuration. In order to avoid confusion, no program example is provided here but programs are widely available in books and periodicals about specific computer systems.

## A Dual Reaction Timer

Studies of competitiveness, comparisons between subjects under identical or nearly identical conditions, or simple sport competition may require that two events or subjects be timed simultaneously relative to a common stimulus. Such a measurement task is analogous to a sport race in which the common stimulus is a starting gun. A dual or multiple timer can of course be obtained by building any of the previously described timing circuits, and this is often the best solution, but timing by means of computer software may be more economical for some applications. The following BASIC program provides a simple reaction timer for two simultaneous subjects. It will run on an Apple II+ or IIe computer and uses the Apple game paddles as input devices.

```
10   GR:COLOR=11
20   FOR I=1 TO 2000
30   NEXT I
```

```
 40   X=0:Y=0
 50   PLOT 15,15
 60   IF PEEK(49249)>127 THEN 100
 70   IF PEEK (49250)>127 THEN 130
 80   X=X+1:Y=Y+1
 90   GOTO 60
100   IF PEEK(49250)>127 THEN 160
110   Y=Y+1
120   GOTO 100
130   IF PEEK(49249)>127 THEN 160
140   X=X+1
150   GOTO 130
160   PRINT"SUBJECT #1 = '';X
170   PRINT"SUBJECT #2 = '';Y
```

After a delay of about 2 seconds this program will plot a red square near the center of the screen. It will then measure the time interval until push button #0 and #1 are pushed. When both buttons have been pushed, the elapsed time for each button is displayed in loop cycles. Each loop cycle represents approximately .0139 seconds.

## A Choice Reaction Timer

A choice reaction timer measures reaction time in which the subject must decide whether to respond or not depending upon the stimulus. For instance, the subject may be asked to press a key as quickly as possible if a green light appears but to ignore a red light, to respond to a green light in a particular location only, or to respond only when a light and a sound occur together. The following BASIC programs will run on an Apple II+ or compatible computer and provide a variety of choice reaction time experiments.

## Red/Green Choice Reaction Timer

```
10   GR
20   FOR I=1 TO 2000
30   NEXT I
40   Y=0
50   COLOR=INT(RND(1)+11.5)
60   PLOT 15,15
70   IF PEEK(49152) THEN 90
80   Y=Y+1:GOTO 70
90   PRINT Y:POKE 49168,0
```

After a delay of about 2 seconds, program line 50 will randomly generate the number 11 or 12, which are the colors pink and green in the computer's low resolution graphics display. A small square in the color selected will be displayed near the center of the screen. The program will then enter a timing loop until any key is pressed. When a key is pressed, the elapsed time expressed as loop cycles will be printed on the screen. The cycle frequency is about 72 per second, which means one cycle takes about .0139 seconds.

## 2-Position/2-Color Choice Reaction Timer

```
10   GR
20   FOR I=1 TO 2000
30   NEXT I
40   Y=0
50   COLOR=INT(RND(1)+11.5)
60   X=INT(RND(1)+.5)*20
70   PLOT X,15
80   IF PEEK (49152)>127 THEN 100
90   Y=Y+1:GOTO 80
100  PRINT Y:POKE 49168,0
```

After a delay of about 2 seconds this program plots a red or green square in one of two locations on the screen and then measures the time interval until a key is pressed. The elapsed time is expressed in loop cycles that occur at the rate of about 72 cycles per second.

## Random Location Reaction Timer

```
10   GR
20   FOR I=1 to 2000
30   NEXT I
40   Z=0
50   X=INT(RND(1)*41)
60   Y=INT(RND(1)*41)
70   PLOT X,Y
80   IF PEEK(49152)>127 THEN 100
90   Z=Z+1:GOTO 80
100  PRINT Z:POKE 49168,0
```

After a delay of about 2 seconds this program plots a red square in a random location on the screen. It then measures the time interval until a key is pressed. The time interval is expressed in loop cycles that occur at the rate of about 72 cycles per second.

## 3-Color Choice Reaction Timer

```
10   GR
20   FOR I=1 TO 2000
30   NEXT I
40   Z=0
50   COLOR=INT(RND(1)*3)+13
60   PLOT 15,15
80   IF PEEK(49152)>127 THEN 90
80   Z=Z+1:GOTO 70
90   PRINT Z:POKE 49168,0
```

After a delay of about 2 seconds this program plots a yellow, blue, or white square near the center of the screen. It then measures the interval until a key is pressed. The time interval is expressed in loop cycles that occur at the rate of about 72 cycles per second.

## 3-Color/3-Location Choice Reaction Timer

```
10  GR
20  FOR I=1 TO 2000
30  NEXT I
40  Z=0
50  COLOR=INT(RND(1))*3+13
60  X=INT(RND(1)*3+1)*10
70  PLOT X,15
80  IF PEEK(49152)>127 THEN 100
90  Z=Z+1:GOTO 80
100  PRINT Z:POKE 49168,0
```

After a delay of about 2 seconds this program plots a yellow, blue, or white square in one of three locations on the screen. It then measures the interval until a key is pressed. The time interval is expressed in loop cycles that occur at the rate of about 72 cycles per second.

## Program Changes for High Precision Timing

```
11  POKE 768,169:POKE 769,128:
    POKE 770,160:POKE 771,0:
    POKE 772,162:POKE 773,0:
    POKE 774,232:POKE 775,240:
    POKE 776,12:POKE 777,205

12  POKE 778,0:POKE 779,192:
    POKE 780,176:POKE 781,248:
    POKE 782,142:POKE 783,46:
    POKE 784,3:POKE 785,140

13  POKE 786,47:POKE 787,3:
    POKE 788,96:POKE 789,9:
    POKE 790,76:POKE 791,9:
    POKE 792,3

XX CALL 768

YY PRINT PEEK(815)*256+PEEK(814)
```

The precision of any of the above programs may be greatly increased with this machine language patch. Lines 11, 12, and 13 should be added and lines XX and YY should replace the last 3 lines of the completely BASIC programs (the IF, LET, PRINT, and POKE 49168 statements). Each program will work as described but the elapsed time will be expressed in loop cycles that occur at the rate of 102,300 cycles per second. The timing performed will be very accurate for purposes of comparison among subjects or trials. For absolute timing purposes, however, an error of up to $+1000$ loop cycles is introduced by the time the computer requires to update its video display. For total times exceeding a bit more than .6 seconds (65536 loop cycles), the count will reset to zero and must start over.

## A Coincidence Anticipation Timer

Given a series of stimuli separated by equal intervals, or intervals that change according to a definite progression, how well can a subject anticipate when the next stimulus will occur? This question is answered by a coincidence anticipation timer. Coincidence anticipation is also the problem to be solved in the familiar "shoot the moving ducks" arcade game. Because a somewhat complex display forms the stimulus in such timers, the use of computer software is the simplest and most versatile means of building one. The following BASIC programs provide a variety of coincidence anticipation timers using an Apple II+ or IIe computer.

## Lengthening Line Coincidence Anticipation Timer

```
10   GR:COLOR=11
20   HLIN 1,19 AT 35:
     HLIN 21,39 AT 35
30   FOR I=1 TO 35
40   FOR J=1 TO 10
50   NEXT J
60   PLOT 20,I
70   NEXT I
80   POKE 49168,0
90   IF PEEK(49152)>127 THEN 110
100  X=X+1:GOTO 90
110  TEXT:HOME:PRINT X
```

The starting display with this program is a horizontal line with one missing element near the bottom of the screen. A vertical line is then slowly drawn from the top down toward that missing element. The experimental task is to press any key as quickly as possible *after* the horizontal line is completed. Any premature keypress is ignored. The total elapsed time in loop cycles occurring at the rate of about 72 per second is displayed after timing is completed.

## Dropping Light Coincidence Anticipation Timer

```
10   GR:COLOR=11
20   HLIN 1,19 AT 35:
     HLIN 21,39 AT 35
30   FOR I=5 TO 35 STEP 5
40   FOR J=1 to 200
50   NEXT J
60   PLOT 20,I
70   NEXT I
80   POKE 49168,0
90   IF PEEK(49152)>127 THEN 110
100  X=X+1:GOTO 90
110  TEXT:HOME:PRINT X
```

This program is identical to the previous one except that the stimulus is a series of squares that appear in descent toward the base line, rather than a solid descending line. Changing the values in line 30 will change the number and separation of the descending stimuli. Changing the terminal value of the loop controlled by line 40 will change the interval between stimuli.

## Visual and Audible Stimulus Coincidence Anticipation Timer

```
10   GR:COLOR=11
20   HLIN 1,19 AT 35:
     HLIN 21,39 AT 35
30   FOR I=5 TO 35 STEP 5
40   FOR J=1 TO 200
50   NEXT J
60   PLOT 20,I
65   PRINT CHR$(7)
70   NEXT I
80   POKE 49168,0
90   IF PEEK(49152)>127 THEN 110
100  X=X+1:GOTO 90
110  TEXT:HOME:PRINT X
```

This program is the same as the previous two except that line 65 has been added. The character string number 7 is a Control-G character which the Apple and other computers recognize as a "Bell" character. Printing a Control-G causes the computer's speaker to beep. The appearance of the lighted square on the screen and the beep are not actually simultaneous but the interval between them is quite small relative to human reaction times.

# Logic Devices and Computer Magic  6

Logic decision-making is a task not often performed electronically by those in the sport sciences but for which digital electronic devices are well suited. Logical decision-making means evaluating the electronic state (on or off) of one or more inputs and setting the state of an output accordingly. Some examples of logical decision-making devices are burglar and smoke alarms, low or high pool water alarms, order of finish detectors, and electronic door locks. Additionally, logic circuits are often integral parts of all kinds of electronic devices; such integrated circuits as counters, decoders, timers, memories, and even microprocessors are actually arrays of simple logic devices, sometimes thousands of them on a single silicon chip.

In the mid-19th century a mathematician named George Boole invented a system of algebra based on the binary number system for the study of logic. His purpose was to allow problems of logic to be precisely stated and so reduced to a mathematical problem. He used the value 1 to mean "true" and the value 0 to mean "false," and devised several operations (addition, subtraction, etc., are operations in traditional arithmetic) that could be used to create logical statements such as "If A is true and B is true, then C is true." Though Boolean algebra was developed for use in philosophy, it happens to be extremely useful in digital electronics which makes wide use of binary mathematics. A full understanding of Boolean algebra is helpful but not essential to understanding electronic logic. Those desiring further study will find chapters on Boolean algebra in many college texts on mathematical logic, general algebra, or computer science.

## Electronic Logic Devices

Electronic logic devices have functions based on the operations of Boolean algebra. For purposes of naming the functions performed, the Boolean "true" state is defined in most devices as a high or positive voltage state. The Boolean "false" state is defined as the ground potential or low state. Though a few devices are made in which this "positive logic" system is reversed, any electronic function produced by a negative logic device can be duplicated by a positive logic device with a different Boolean function name. Some common logic devices in integrated circuit form are the inverter, noninverting buffer, AND gate, OR gate, EOR gate, NAND gate, NOR gate, Exclusive NOR gate, and a few compound function gates. The symbol and truth table of the simplest form of each logic device is presented in Figure 6.1, and the operation of each is as explained briefly below. The integrated circuit numbers of a wide variety of these and other devices is given in Appendix B.

**Inverter.** An inverter performs the Boolean NOT function which may be stated as, "If A is true, then B is false." An inverter always has only one input and its output is always in the state opposite the input.

**Noninverting Buffer.** A noninverting buffer has one input and one output, and the output is always in the same state as the input. Though the function performed is the same as that of a piece of wire, the noninverting buffer does serve a purpose. In situations in which a voltage drop occurs because of long wires or any other reason, noninverting buffers can be used as boosters much like telephone line repeaters.

**AND Gate.** The AND gate performs a Boolean AND operation which may be stated as, "If A is true and B is true, then C is true." AND gates may have two, three, four, or even more inputs, but only one output. The output will be high when *all* inputs are high, and low if any inputs are low.

**OR Gate.** The OR gate performs a Boolean OR operation which may be stated as, "If A is true or B is true, then C is true." OR gates may have two or more inputs but only one output. The output will be high if either or both inputs are high, and low when both inputs are low.

**EOR Gate.** The EOR or Exclusive OR gate performs an operation which may be stated as, "If A *or* B is true, but not both, then C is true." An EOR gate may have two or more inputs but only one output, and that output will be high if any one input is high but will be low if more than one input is high or if all inputs are low.

**NAND Gate.** The NAND gate performs the operation of an AND gate with an inverter attached to its output. The output of a NAND gate is low when all of the inputs are high and will be high if any input is low.

**NOR Gate.** The NOR gate performs the operation of an OR gate with an inverter attached to its output. The output of a NOR gate will be low when any or all inputs are high and will be high when any of the inputs are low.

**Exclusive NOR Gate.** The exclusive NOR gate performs the operation of an EOR gate with an inverter attached to its output. The output will be low when any of the inputs is high, but will be high if all inputs are low or if more than one input is high.

**Compound Function Gates.** Integrated circuits with compound designations such as AND-OR, or AND-OR-INVERT, perform the designated operations in sequence. An AND-OR gate, for example, performs the AND operation

**Inverter / NonInverting Buffer**

| IN | OUT | |
|---|---|---|
| | Inverter | NonInverting Buffer |
| O | I | O |
| I | O | I |

**OR / NOR**

| IN | | OUT | |
|---|---|---|---|
| A | B | OR | NOR |
| O | O | O | I |
| O | I | I | O |
| I | O | I | O |
| I | I | I | O |

**AND / NAND**

| IN | | OUT | |
|---|---|---|---|
| A | B | AND | NAND |
| O | O | O | I |
| O | I | O | I |
| I | O | O | I |
| I | I | I | O |

**EOR / ENOR**

| IN | | OUT | |
|---|---|---|---|
| A | B | EOR | ENOR |
| O | O | O | I |
| O | I | I | O |
| I | O | I | O |
| I | I | O | O |

**Figure 6.1** Symbols and truth tables for some common logic devices.

on each of two or more groups of inputs and then performs the OR operation on the outputs from the AND operations. Such ICs are used more efficiently with available current than are multiple ICs, and they greatly simplify the construction process.

## A Majority Vote Detector for Weightlifting

In the sport of Olympic weightlifting, a panel of three judges must certify that a lift has been made within the rules. For a lift to be accepted, at least two of the judges must agree that it was valid. If the judges are designated as "A," "B," and "C," the requirement for acceptance of a lift may be written as, "If A and B vote yes, or if A and C vote yes, or if B and C vote yes, then the lift is good." If each judge is to signify a proper lift by closing a switch, a good

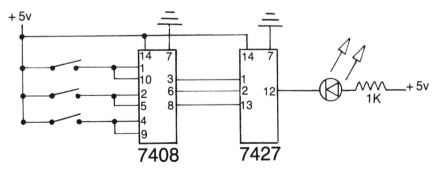

**Figure 6.2** The LED will be illuminated when any two of the switches are closed. Note that though the LED is driven direct- ly in this circuit, higher current devices would have to be isolated by a relay to avoid burning up the circuit.

lift signal can be constructed by ANDing the wires from each pair of judges and then NORing the results and using the final output to activate a signaling device such as a light. The required circuit is shown in Figure 6.2

## A Touch Detector for Fencing

Electric sensing is used in fencing with the foil and the epee to detect scores. When the jacket of one fencer is touched by the opponent's weapon in a place that constitutes a legal hit, a light is illuminated to signal the hit. But as a hit is recorded, the detection system is deactivated and subsequent hits are not detected. The circuit shown in Figure 6.3 is for such a device. The two switches each represent the weapon and jacket of opposing fencers. When one switch is closed because a hit is scored, one of the monostable multivibrator circuits will be triggered and will light an LED to signal the hit. At the same time and for the duration of the monostable's pulse (about 2 seconds with the resistor and capacitor values shown), the output of the 7400 NAND gate controlling the circuit to detect hits against the fencer scoring the first hit is deactivated. In a device

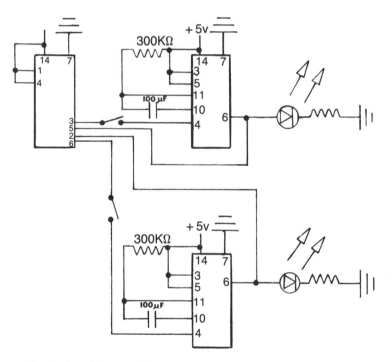

**Figure 6.3** Closing either switch illuminates one of the LEDs and disables the other switch for 2 seconds. Higher current signaling devices must be isolated to avoid burning up the circuit.

designed for actual use, the circuit through each LED would probably be replaced by a relay circuit in order to drive a signaling device requiring more current than an LED.

## Security Alarms

It is essential at many kinds of sport facilities to maintain strict control of access. This is especially important for swimming pools, rooms containing gymnastics equipment, and other such areas where the need to protect the facility is superseded by the need to prevent accidents caused by unauthorized access. Security can be enhanced for such facilities by electronic monitoring of doors.

Monitoring the status of the doors to a facility requires employing a switch that can sense the status (e.g., open vs. closed, locked vs. unlocked) of each door, as well as some simple logic circuitry to control indicator lights, buzzers, or alarms. Because contact bounce is not likely to be a problem even if it is severe, any kind of switch whose output follows the status of the monitored door may be used.

One option for signaling or indicating an open door is to use a circuit that remains closed as long as the door remains closed but that will open if a door is opened, using the current flowing through the circuit to light an LED or light to signify that the door is secure. As shown by Figure 6.4, several such circuits can be used as inputs to a NAND gate whose output controls an alarm to give warning if any door is opened.

Another approach is to use circuits that are normally open but that will be closed if a door is opened, as illustrated in Figure 6.5. Here the LEDs are illuminated to signal which door has been opened and an OR gate is used to trigger the alarm. This method is more suitable for battery operation because significant current is drawn only when a door is opened; however, a dead battery will defeat the purpose of the detection system. Note that the circuits shown in Figures 6.4 and 6.5 will activate the alarm only while the door is actually open.

**Figure 6.4** All four switches are normally left closed. If any switch is opened, the corresponding LED will light and the alarm will be triggered.

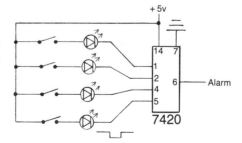

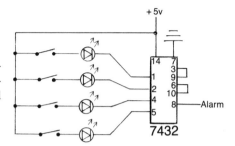

**Figure 6.5** All four switches are normally left open. If any switch is closed, the corresponding LED will light and the alarm will be triggered.

Other detecting systems and alarms can be constructed following the above examples by modifying the trip mechanisms or switches. The square wave symbol in each diagram shows the output of the switch when it is tripped.

Figure 6.6 shows a device for sensing unauthorized or after-hours activity in a swimming pool by detecting wave action. A mercury tilt switch is mounted in a float so that any rocking of the float will momentarily close the switch and trip the alarm.

Figure 6.7 shows how an interrupted light beam can be used to detect an intrusion. Connected to a counter rather than to an alarm, such a device might also be used at a turnstyle as an attendance meter.

A pressure switch in a floor plate, as shown in Figure 6.8, can also be used to detect the passage of people or vehicles. Two such circuits could be used to start and stop a timer in experiments measuring jumping power.

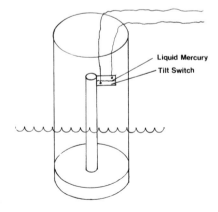

**Figure 6.6** The mercury tilt switch contains a small quantity of liquid mercury which covers or uncovers the contacts according to the position of the switch. Do not try to construct such a switch from scratch; mercury is a toxic substance.

**Figure 6.7** The voltage through the photoresistor drops when the light beam is interrupted. The 7414 inverts the signal and conditions it for use in digital circuitry. A buffer or second inverter could be used if a low output were needed.

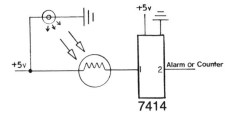

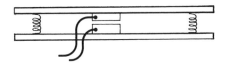

7414

**Figure 6.8** The switch is closed when an object of sufficient weight to compress the springs is placed on the upper surface.

## An Automatic Light Switch

Tennis courts and other outdoor facilities with evening hours of operation, passageways between buildings, parking lots, and similar areas are always safer when well illuminated. Outdoor lighting is usually controlled by hand-operated switches, which means someone must remember to turn them on. Also, such lights can easily be turned off by unauthorized persons. Both problems can be solved with an automatic switch, although a simple timer switch is not the complete solution because the sun sets at varying times throughout the year. By using a circuit that goes high in response to the lack of light striking a photoresistor, and ANDing the output with the output of a timer circuit that goes high between 4:30 p.m. and 1 a.m. (see Figure 6.9) to control the lights, one can make the lights come on if it is after 4:30 p.m. and it is dark, and to go off precisely at 1 a.m.

## Computer I/O Boards

An I/O board for a computer is an interface circuit that allows the computer to read digital data from the outside (see the input board described in chapter 3) and to send digital data to the outside. Such boards are readily available as attachments for most of the popular microcomputers, and some computer models even have I/O capabilities built in. I/O boards transmit data in parallel format; that is, a data word of eight or more bits is transferred to or from the microprocessor simultaneously rather than in sequence. Under control of the BASIC language, data is read into the microprocessor using the PEEK statement and sent by the microprocessor using the POKE statement. Between reading and writing of data, the microprocessor can be used to perform all kinds of complex logic or other operations such as the permanent saving or display of data.

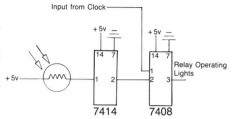

**Figure 6.9** The output of the 7414 is high when insufficient light strikes the photoresistor. The input to the 7408 from the clock is also high between 4:30 p.m. and 1 a.m. The output from the 7408 is high when both inputs are high.

An I/O board is not like a parallel interface board. With a parallel interface board, several control lines are provided to the outside world in addition to the data lines and these are used to control the sequential transmittal of more than one byte or data word. The BASIC commands used for input and output using a parallel interface are INPUT, GET, and PRINT.

Though it is usually simpler and less expensive to purchase an I/O board than to build one, a simple I/O board can be constructed fairly easily. The circuit shown in Figure 6.10 is designed for use with an Apple II+ computer but can be used with any computer that allows access to the data bus, four least-significant address lines, a control line representing the 4k block of memory used for I/O purposes, and the read/write control line.

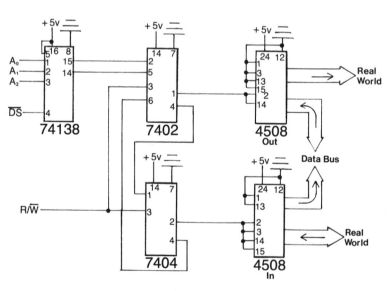

**Figure 6.10** Note that the input and output operations here are not mirror images of each other. In the input function, data reaches the outputs of the latch (and the computer data bus) only when the input port is addressed in a PEEK statement (or the equivalent in other languages). In the output function, data reaches the latch outputs (and the outside world) all of the time and is only changed when the output port is addressed.

This circuit uses separate 8-bit ports for input and output and different addresses to access each port. Whenever this or any other IO board is used to drive (provide power to) any external device, it is important that the current drawn by the device does not exceed the capacity of any part of the interface or of the computer to supply current. Also, it is important to provide cooling ventilation to the controlling computer.

## An Order of Finish Detector

Determining the order in which several events occur is a task required in all racing sports. Usually this is done by a panel of judges observing the finish, the use of photography at the moment the finish line is crossed, and the comparison of race times using separate but synchronized timers for each competitor. An order of finish detector for up to eight competitors can be made using a computer equipped with an 8-bit I/O board. The circuitry required is very simple. A set of eight switches, which are closed at the moment each competitor finishes, must be connected to the input port of the I/O board so that each competitor is represented by one bit of the input byte. Each input bit will then go high when the corresponding competitor finishes the race; the process of sensing and interpreting the changes in the input bits is handled by software.

For most applications of this type, the order of finish program would be written in assembly language in order to maximize the speed of execution and improve the device's limit of resolution. To make the process easier to comprehend, however, the order of finish program which follows is written in BASIC. The program is designed for an Apple II+ or IIe computer with the I/O board placed in peripheral slot number 4. With the addition of a real time clock board in another peripheral slot, only a slight program modification is required to obtain times for each competitor as well as the order of finish.

```
10   HOME:X=0:A=1
20   Y=PEEK(49343)
30   IF Y-X<>0 THEN 50
40   GOTO 20
50   Z=ABS(X-Y)
60   IF Z=1 THEN L(1)=A
70   IF Z=2 THEN L(2)=A
80   IF Z=4 THEN L(3)=A
90   IF Z=8 THEN L(4)=A
100  IF Z=16 THEN L(5)=A
110  IF Z=32 THEN L(6)=A
120  IF Z=64 THEN L(7)=A
130  IF Z=128 THEN L(8)=A
140  X=Y:A=A+1:IF A=9 THEN 160
```

After clearing the screen and initializing some values, this program repeatedly reads the memory location containing data from the sensors. After each read it compares the data with the previous value to see if a change has occurred and, if it finds a change, the program jumps to line 50 to determine which switch has been closed. After all eight switches have been detected as closed, a table is printed showing the order in which each was closed. For correct operation it is essential that each switch remain closed once it has been closed. An assembly language program

```
150   GOTO 20
160   PRINT"LANE PLACE"
170   FOR I=1 TO 8
180   PRINT" ";I;" ";L(I)
190   NEXT I
```

to perform the same function could do so in a way that would remove the necessity of each switch remaining closed.

## An Environmental Control System

Using a general purpose microcomputer for sense and control applications such as access limitation, thermostatic control, or monitoring of lighting conditions is a bit like using a fire hose to extinguish a match. It is usually easier and cheaper to use single purpose, dedicated devices such as the common thermostat or, in more complex situations, to use specially designed microprocessor controlled devices for such applications. However, the requirements of such a system are similar whether used for controlling a building's environment or a set of experimental apparatus in a laboratory such as the data collection devices used in a stress test.

The following BASIC program illustrates how a control system might be designed. The program would work with up to eight environmental sensing devices, each of which may be on or off to indicate two different conditions, and up to eight output devices of the same type. As written, the program will turn the output devices on or off to match the input from the corresponding input devices. The program causes the computer to continuously check the inputs searching for any changes. When a change is detected, the program jumps to a subroutine that calculates which input bit has changed and then responds by setting the appropriate output bit. Rather than simply setting an output bit, the computer could be programmed to modify the output depending upon the time of day, weather conditions, traffic flow measurements, or other factors. The computer could also store or print raw or analyzed data on a regular basis.

```
10   HOME:X=0
20   Y=PEEK(49343)
30   IF Y-X<>0 THEN 50
40   GOTO 20
50   POKE 49344,Y
60   X=Y
70   GOTO 20
```

This program repeatedly reads the memory location containing data from the sensors. After each read it compares the data with the previous value to see if a change has occurred. If it finds a change, the program jumps to line 50 which pokes a new value into the memory location controlling the output port. The new value is then substituted for the old, and the scanning process is restarted. For a more complex output requirement, line 50 would be replaced by a GOSUB statement to the necessary subroutine.

## A Memory Drum

A memory drum, a well established experimental tool in the study of verbal learning, can present a series of stimuli consisting of printed text in a controlled order. The sport scientist might use such a device to help evaluate aptitude for such sport skills as learning team sport signals or as a novel task in studies of effects of environmental stress in activities such as scuba diving or mountaineering. Most memory drums are somewhat primitive mechanical devices consisting of a mechanism for moving a sheet of paper behind a viewing window so that only one line of text may be read at a time. The device may or may not automatically move the paper or measure or limit the viewing time.

Several types of experiments employ the memory drum. In serial learning studies, the subject is presented with a series of nonsense words or syllables and asked to memorize them in sequence. He or she is then asked to recite the list, with feedback given after each word. In paired-associate learning studies, the subject is asked to memorize pairs of words or syllables and then, given a stimulus of one word of a pair, to recite the other. In some such studies the pairs are presented one at a time during the learning process, and in others they may be presented together.

Using a microcomputer in place of a mechanical memory drum offers considerable advantages of flexibility, precision of measurement of timing and, assuming the computer is already available, of cost. The following BASIC programs provide examples of several kinds of memory drum experiments performed entirely by a computer. They are written to run without modification on an Apple II computer and should be easily adaptable to any computer using the BASIC language.

## Serial Learning Task

```
10   HOME
20   PRINT"PROGRAMMING SEQUENCE"
30   FOR I=1 TO 5
40   PRINT"ENTER WORD # ";I;" ";
50   INPUT WD$(I)
60   NEXT I
70   INPUT"ENTER VIEWING
     TIME IN SECONDS";T
80   HOME:PRINT"PRESS ANY KEY
     TO BEGIN":GET A$
90   FOR I=1 TO 5
100  HOME:PRINT WD$(I)
110  FOR J=1 TO T*1000
120  NEXT J:NEXT I
```

This program guides the experimenter through entry of a sequence of five words or syllables to be memorized, and the viewing time for each word. It then prompts the subject to enter each word in the sequence while providing accuracy feedback. The program also automatically tallies the number of errors made by the subject. For brevity the number of words is limited to five, though in actual practice a larger number would be used.

```
130  HOME:ER=0
140  PRINT"ENTER THE WORDS
     ONE AT A TIME. PRESS THE
     RETURN KEY AFTER EACH WORD."
150  PRINT"THE CORRECT WORD
     WILL BE PRINTED AFTER YOUR
     CHOICE IS ENTERED."
160  FOR I=1 TO 5
170  PRINT"ENTER NUMBER ";I;" ";
180  INPUT RS$(I)
190  PRINT"THE WORD WAS ";WD$(I)
200  IF RS$(I)<>WD$(I) THEN ER=ER+1
210  NEXT I
120  PRINT"TASK COMPLETED
     WITH ";ER;"ERRORS."
```

## Paired Associate Learning Task

```
10   HOME
20   PRINT"ENTER LIST"
30   FOR I=1 TO 5
40   PRINT "ENTER PAIR ";I;"
     SEPARATED BY COMMA"
50   INPUT WD$(I, 1),WD$(I,2)
60   NEXT I
70   INPUT"VIEWING TIME";T
80   HOME: PRINT"PRESS ANY KEY
     TO BEGIN":GET A$
90   FOR I=1 TO 5
100  HOME:PRINT WD$(I,1)
110  FOR J=1 TO T-1000:NEXT J
120  NEXT I
130  HOME:ER=0
140  PRINT"PRESS ANY KEY TO
     BEGIN":GET A$
150  FOR I=1 TO 5
160  PRINT WD$(I,1)
170  INPUT RS$(I,2)
180  PRINT"THE WORD WAS ";
     WD$(I,2)
```

This program guides the experimenter through the entry of a set of five pairs of words and adjusts the time allowed for viewing each pair during the memorization process. After each word in the list is entered, the program displays each pair of words for the specified time. The subject is then shown one word from each pair and asked to enter the corresponding word. The correct word is displayed after each entry and a tally of errors is printed before the program ends.

```
190  IF RS$ (I,2)<>WD$(I,2)
     THEN ER=ER+1
200  NEXT I
210  PRINT"TOTAL ERRORS= ";ER
```

## A Delayed Knowledge of Results Feedback Device

Experiments in which results feedback is manipulated clearly show that performance suffers when feedback is delayed, withheld, or interfered with in any way. A typical experiment may require the subject to try moving a marker a specified distance along a track or to rotate a dial through a certain number of degrees. Delayed feedback regarding accuracy or speed of movement is provided after a specified interval following completion of the task. Many simple but dynamic feedback experiments may be performed using a simple videogame format with the game adapted to provide for various feedback conditions and for scoring. The game itself of course becomes a possible performance factor, as it provides an element of motivation that may not be present in more traditional devices. Several delayed feedback experiments are provided in the following BASIC programs.

## Graphic and Verbal Knowledge of Results

```
10   HOME:GR:COLOR=10
20   HLIN 0,39 AT 21:PLOT 20,
     22:PLOT 22,22
30   IF PEEK(49249)<128 THEN 30
40   FOR I=1 TO 2000:NEXT I
50   PRINT CHR$(7)
60   FOR I=1 TO 1500:NEXT I
70   X=INT(PDL(0)/7)
80   PLOT X,20
90   ER=ABS(21-X):PRINT"ERROR
     = ";ER
100  IF PEEK(49249)=>128 THEN 10
110  GOTO 100
```

This program explores the effects of postknowledge of results on accuracy of movement in a dial turning task. The subject is presented with a horizontal track and a target. The task is to move a ball along the track to the target within a specific time period. The moving ball, however, is not visible to the subject until the movement time has elapsed. The subject signals readiness by pressing the game paddle button. After about 2 seconds the computer signals the beginning of the movement period with a beeping sound and the subject may turn the dial of the game paddle. After about 1 second the movement period ends, the ball appears, and the amount of error is displayed. A press of the game paddle erases the screen and restarts the program.

## Verbal-Only Knowledge of Results

```
10   HOME:GR:COLOR=10
20   HLIN 0,39 AT 21:PLOT
     20,22:PLOT 22,22
30   IF PEEK(49249)<128 THEN 30
40   FOR I=1 TO 2000:NEXT I
50   PRINT CHR$(7)
60   FOR I=1 TO 1500:NEXT I
70   X=INT(PDL(0)/7)
80   ER=ABS(21-X):IF ER=0 THEN 120
90   IF X>21 THEN 110
100  PRINT"LEFT BY ";ER:GOTO 130
110  PRINT"RIGHT BY ";ER:GOTO 130
120  PRINT"DIRECT HIT"
130  IF PEEK(49249)=>128 THEN 10
140  GOTO 130
```

This program is very similar to the previous one except that the graphic display of the moving ball is replaced with the strictly verbal information of the deviation to the left or right of the target.

## No Knowledge of Results

```
10   HOME:GR:COLOR=10:ER=0:
     FOR TR=1 TO 10
20   HLIN 0,39 AT 21:PLOT
     20,22:PLOT 22,22
30   IF PEEK(49249)<128 THEN 30
40   FOR I=1 TO 2000:NEXT I
50   PRINT CHR$(7)
60   FOR I=1 TO 1500:NEXT I
70   X=INT(PDL(0)/7)
80   ER=ABS(21-X):TE=TE+ER
90   IF PEEK(49249)=>128 THEN 110
100  GOTO 90
110  NEXT TR
120  PRINT TE
```

This program is the same as the previous two but no knowledge of results of any kind is provided. The process of attempted positioning of the moving ball (invisible) is repeated 10 times and then the cumulative error (in absolute terms) is displayed. Before this program is used, the subject should be given a chance to learn the movement pattern by using one of the other programs. The concurrent feedback program below is ideal for this purpose though the movement time (the 1-to-35 loop) may need adjustment.

## Concurrent Feedback

```
10   HOME:GR:COLOR=10
20   HLIN 0,39 AT 21:PLOT
     20,22:PLOT 22,22
```

This program provides a task similar to the others given here but provides visual concurrent feedback of results. How-

```
30   IF PEEK(49249)<128 THEN 30
40   FOR I=1 TO 2000:NEXT I
50   PRINT CHR$(7)
60   FOR I=1 TO 35
70   X=INT(PDL(0)/7)
80   COLOR=0:PLOT Y,20:
     COLOR=10:Y=X
90   PLOT Y,20
100  NEXT I
110  ER=ABS(21-X):PRINT"ERROR = ";ER
120  IF PEEK(49249)=> 128 THEN 10
130  GOTO 120
```

ever, the subject is required to perform the task in a very limited period of time. This period is 35 cycles of the loop started by line 60. The period can be changed by changing the ending value of I in that loop or it can be made flexible by replacing the 35 with a variable name and adding an INPUT statement earlier in the program to define the variable.

## Delayed Feedback

```
10   DIM X(100):Z=0:INPUT
     "DELAY ?";D
20   HOME:GR:COLOR=10
30   HLIN 0,39 AT 21:PLOT
     20,22:PLOT 22,22
40   IF PEEK(49249)<128 THEN 40
50   FOR I=1 TO D:X(I)=0:
     NEXT I
60   FOR I=1 TO 2000:NEXT I
70   PRINT CHR$(7)
80   FOR I=1 TO 75
90   X(I+D)=INT(PDL(0)/7)
100  IF X(I)<>Y THEN 120
110  GOTO 140
120  COLOR=0:PLOT Y,20:
     COLOR=10:Y=X(I)
130  PLOT Y,20
140  IF Z=1 THEN 180
150  NEXT I
160  FOR I=75 TO 75+D
170  Z=1:GOTO 100
180  NEXT I
190  ER=ABS(21-X):PRINT"ERROR = ";ER
200  IF PEEK(49249)=> 128 THEN 10
210  GOTO 200
```

This program is the same as the concurrent feedback program above except that the ball's movement in response to the subject moving the game dial is delayed by the experimenter's response to the INPUT statement early in the program. The delay factor may be any integer value from 1 to 125, though values above about 25 may make the delay period too long to be practical. Values above 125 will cause an error message and terminate the program.

# Analog to Digital Interfacing

Microprocessors and all other integrated circuit components used in computers, except for a few things like resistors and capacitors, are digital devices. The digital world in which they operate is a simple one where everything is reducible to discrete representation in binary mathematics and there is no such thing as an ambiguity. The real world, however, is an analog world, where things may vary on a continuous scale. Whereas a digital component can recognize only two conditions (high and low), for example in a voltage range of 0 to +5 volts, in the real or analog world there is an *infinite* number of conditions within any voltage range. If a computer or other digital device is to be used to measure any factor or component of the analog world, it must be provided with interfacing circuitry. That circuitry must first convert the real world variable (i.e., pressure, temperature, sound or light intensity, chemical concentration) to an analog or proportional electrical signal or voltage level that is within an acceptable range, and then convert the analog signal to a digital approximation that can be interpreted by the computer.

## Analog to Digital Conversion

Electronically, the conversion of an analog value to a digital one requires some fairly complex circuitry. The usual method is to make successive approximations with precision increasing to the limits of the digital device. A bracketing process is followed in which the digital device repeatedly makes a digital estimate of the unknown voltage, converts the guess to an analog signal with a digital to analog converter, and compares the known value to the unknown.

Fortunately, all of the successive approximation circuitry construction can be avoided through the use of special integrated circuits called analog to digital or A/D converters in which all of the circuitry has been placed on a silicon chip. Input to these chips is an unknown analog voltage (known to be within an acceptable range, however), and output is a digital value carried on a data bus of eight or more data lines. Such ICs can be obtained as components for use in a custom designed interface, or they may already be built into A/D boards for use in specific computers.

## A/D Interface Boards for Computers

Provided one is available that meets the technical requirements for a given application, the easiest and probably least expensive method of giving a computer an analog interface is with an "off the shelf" A/D interface circuit board. Such interfaces are readily available for most of the popular microcomputers on the market, although few retail outlets carry them in stock. One may have to search through the advertising pages of technically oriented computer periodicals such as Byte or phone the manufacturers of devices that appear to meet the proper requirements.

A/D boards will generally plug into one of the peripheral attachment slots or ports of the computer for which it is designed. Once installed, the A/D board is usually designed to appear to the microprocessor as a memory location that can be read or written to like any other memory location. Reading from BASIC is done using the PEEK statement; the LDA, LDX, or LDY statements are usually used from 6502 assembly language. Using a computer with an 8-bit data bus, the most common current configuration, boards providing up to 8 bits of resolution can be read with a single PEEK statement whereas boards having up to 16 bits of resolution can be read with two PEEK statements.

The main technical points to consider when matching an off-the-shelf A/D board to a specific application are the precision of resolution, speed of conversion, number of available channels, and input range. The precision of resolution is measured in bits and generally ranges from 8 to 16. An 8-bit A/D board has 256 different output combinations, whereas a 12-bit board has 4,096 output combinations. Thus, 8- and 12-bit resolution boards with an allowable analog input range of 10 volts will give resolution to the nearest 0.0391 volts and 0.00196 volts, respectively. The speed of conversion is the time required to complete a conversion.

A/D boards vary widely in speed from less than 100 conversions per second to many thousands, depending on the method and technology used. High speed, absolutely essential for some applications but irrelevant in others, is quite a bit more expensive than lower speeds. The number of channels is the number of different analog inputs that can be interfaced through the board at any one time. The input range is the range of analog voltages that can be converted. Any voltage above or below the specified limits will saturate the digital output, leaving it all 1's or 0's, and may burn out the converter. Because the many factors to consider make designing a complete circuit quite difficult until every aspect of a given application is known, only a few complete examples of specific devices will be given here.

**Figure 7.1** Some of the pin designations require explanation. Pin 11 is called "Blank and Not Convert." When it is made high the outputs are disabled. Conversion commences when it is brought low. Pins 13 and 14 are the analog connections. Pin 14 is the analog common connection. Pin 15 is called "Bipolar Off." When connected to the analog common line, the analog input range is 0 to +10. When it is unconnected the range is −5 to +5 volts. Pin 17 is called "Not Data Ready." This line goes low when a conversion is complete and the data is available at the outputs.

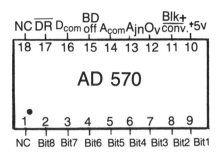

## A Custom A/D Interface Board

When there is no off-the-shelf analog to digital converter board available for a particular computer or application, one may be designed and constructed from scratch. Such a project requires only a few integrated circuits for A/D conversion, latching, and control functions, plus access to the microprocessor data, address, and control lines.

A popular and reasonably low priced A/D circuit is the AD570 produced by Analog Devices (see Figure 7.1 for pinout). The AD570 is a complete 8-bit converter housed in an 18-pin DIP, and requires only 25 microseconds to complete a conversion.

A circuit demonstrating how the AD570 can be used to make an A/D board for an Apple II+ computer and accepting analog input from 0 to +10 volts is shown in Figure 7.2

Because the AD570 requires 25 microseconds to complete a conversion, this interface board requires two program commands separated by at least that much time in order to work properly. Conversion is initiated by a POKE (or write command in other languages) to a valid address. While the conversion is taking place, the AD570's Data Ready line, connected to the clear of the latch circuit, is high and keeps the latch outputs set at zero. When the conversion is completed, the Data Ready line goes low and the latch outputs will match those of the AD570. The PEEK or read command to the same address and at least 25 microseconds after the write will freeze the latch outputs while the data is transferred to the computer.

## An 8-Channel A/D Interface Board

The AD7581 by Analog Devices is an 8-channel, 8-bit A/D converter available in a 24-pin DIP. (See Figure 7.3 for pinout.)

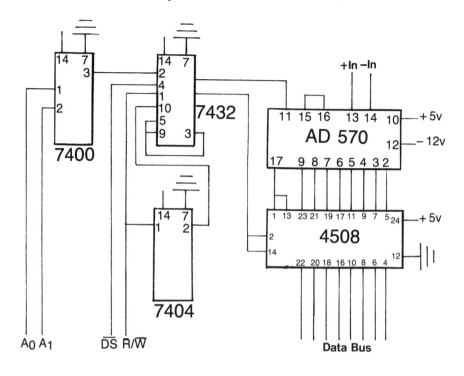

**Figure 7.2** Note that the AD570 requires +5 and −12 volt connections whereas the other circuits operate on +5 and ground connections. The lines labeled A0, A1, Ds, and R/W go to connections in the Apple computer's peripheral slots. The A/D interface would work in any computer having equivalent connections.

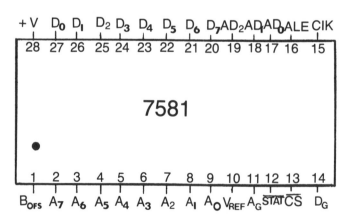

**Figure 7.3** For details on the use of this circuit, the reader is advised to consult the data sheet available from Analog Devices Corp.

Though not as fast in performing a conversion as the AD570, the AD7581 is relatively easy to use, provides eight analog inputs, and is functionally faster than the AD570. It also contains its own built-in address decoding, latching, and timing circuitry which greatly simplifies its use with computers. The circuit given in Figure 7.4 illustrates how this is done. Each of the eight input channels is accessed by a PEEK or read statement to a specific address. With the interface board shown connected to an Apple II+ computer through peripheral slot number 4, the valid addresses would be 49344 to 49351. Because the AD7581 is continuously performing A/D conversions on the inputs to all eight analog inputs in sequence, there is no need to wait for a conversion to be completed, even though the conversion speed is slower than the AD570. A PEEK to any channel's address will obtain the result of the most recent completed conversion.

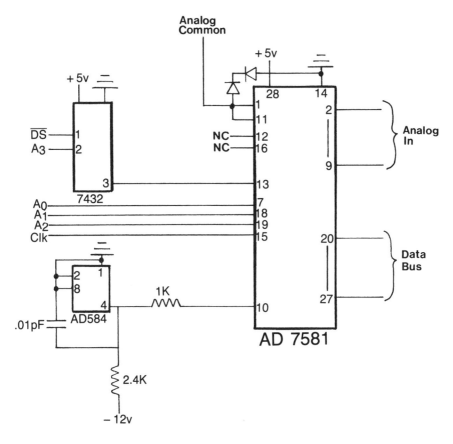

**Figure 7.4** This circuit continuously performs conversions on all eight input channels. Data from any channel appears at the outputs when that channel is addressed in a PEEK statement.

## Transducers

Many of the parameters the sport scientist might wish to measure or record using a computer with an A/D interface, such as temperature, force, weight, velocity, pressure, gas concentration, or light intensity, are not electrical in nature. A transducer must be used to convert the variable of interest to a proportional voltage (or to a current and then to a voltage), which may be used directly or after further conditioning as input to the A/D circuit.

Temperature is an important factor in many sport settings. The exercise physiologist is interested in the temperature of muscle tissue and in core body temperature during exercise. Air temperature is often a factor in athletic performance, and water temperature is critical in aquatic sports.

A thermistor is a simple device whose electrical resistance changes with temperature within a certain range. It also becomes an excellent remote sensing device if the distance between the thermistor and the controlling computer is not too great. By selecting a thermistor that operates with the proper supply voltage, within the temperature range of interest, and which can tolerate the necessary environmental conditions, it is usually possible to connect the thermistor directly to the A/D interface to make a thermometer for nearly any purpose.

Some simple integrated circuits are also available for sensing temperature. Such circuits usually exhibit a high degree of linearity in the relationship of their output to temperature, and they can supply output in the form of a variable current rather than a variable voltage. This feature is essential when the sensor is to be located a long distance from the controlling computer because of the resistance introduced by long connecting wires.

Light intensity may be a factor in any study in which an athlete or experimental subject must detect or track an object, and particularly when colors must be distinguished. It may also be an important factor in the safety of a sports facility such as a swimming pool.

A photoresistor, sometimes called a photo conductive cell or photo cell, changes its resistance according to the intensity of light striking its active surface. As the light intensity increases, the cell's resistivity decreases and the output voltage (assuming a constant input voltage) increases.

A solar cell is a device which generates an electric current when light falls on its active surface. It is important to realize that with a solar cell it is the output current, and not the voltage, that varies with light intensity.

The student of biomechanics is often interested in the measurement of position, as in the measurement of the angle between two bones at a joint or in the location (in Cartesian or polar coordinates) of each major joint when a photograph of a performer in action is projected on a flat surface.

Position can often be converted to an electrical signal using a potentiometer, a mechanical device whose resistance changes according to the position of a movable shaft or slidepiece. Angles may be measured using a rotary or dial type "pot," and position in a linear direction can be measured with a slide or a dial type which is turned by a linear movement using gears or pulleys. One problem that can often be overcome with computer software is that many potentiometers give an output which is not linear in relation to the physical movement measured.

Concentrations of gases such as oxygen or carbon dioxide are frequently measured in exercise research and in assessing fitness levels of athletes. Though special transducers can be acquired for this, they are not very easy to use. However, most exercise laboratories are already equipped with electronic gas analysis apparatus, and though this equipment may not already be interfaced to a computer, it generally can be. At some point in the circuitry of the gas analyzer there will almost always be a voltage level that is proportional to the gas concentration. The analyzers made by the Beckman Company have a built-in access port or plug receptacle to this point. If the voltage range does not exceed the capacity of the A/D circuit, it can be used directly as input.

The measurement of gas or liquid pressure, and of other forms of force such as weight and strain, is required in many aspects of sport science. Such measurements may be computerized through use of special integrated circuits designed for the purpose. The selection of such circuits and the type of package housing them depends upon the range of pressures to be encountered and the type of gas or liquids involved, in addition to the electrical requirements of the A/D interface. Data books of various circuit manufacturers should be consulted for complete specifications and suggestions for practical applications.

## Signal Amplification

Many analog voltage signals encountered in sport science applications, whether natural or derived from various transducers, are frequently too small to be directly used as input to an A/D interface circuit. Even when the voltages obtained from a transducer are large enough to be read, the changes in voltage that occur with changes in the measured parameter (and it is the change that is of interest) are frequently too small. A typical 8-bit A/D circuit, for example, has an operating input range of 10 volts which gives it a resolution of slightly better than .04 volts, whereas the voltage changes produced by a typical temperature transducer such as the LM334 are on the order of 10 millivolts per degree. In such cases an amplifier may be needed to increase the amplitude of the signal to a level within the input range of the A/D circuit, but using as much of that range as possible.

An Op-Amp is a single and inexpensive integrated circuit which can be used to make a wide variety of amplifying circuits. Amplitude gains of as much as X100,000 can be obtained using an Op-Amp and a few common resistors, though gains of from X10 to X1000 are more practical. Figure 7.5 shows several amplifying circuits using the common and popular 741 Op-Amp.

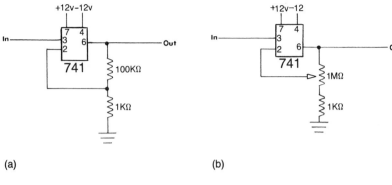

(a)                              (b)

**Figure 7.5** (a) A X100 noninverting dc amplifier. (b) An X1 to X100 variable noninverting dc amplifier. (c) An X10, X100, X1000, and X10000 switch selectable noninverting amplifier. The gain of the amplifier is determined by the ratio of the two resistors.

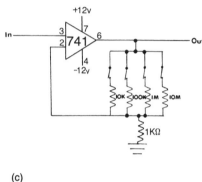

(c)

## Temperature Biofeedback

Biofeedback is apparently useful in the treatment of various medical problems and in relaxation training, and has been used to help athletes learn to reduce tension. Evidence suggests that such training may improve performance in activities such as basketball foul shooting, where excess tension levels may interfere with concentration. Physiologic functions that are normally considered beyond volitional control are detected with electronic apparatus and fed back, or displayed, to the subject in some meaningful manner.

Temperature biofeedback is usually in the form of a series of lights and/or tones that are activated as the temperature of a finger or other body part varies. A computer-controlled temperature biofeedback device can be made using an A/D circuit to operate a thermistor in a bridge amplifier circuit (see Figure 7.5) or other temperature sensing element that can operate in the appropriate temperature range. With the thermistor in contact with the subject's finger, the computer can then provide a feedback display that is more meaningful to the subject than that provided by other devices. The following programs demonstrate how the Apple computer's graphics can be used to display a thermometer that rises and falls with the temperature. So that readers without the temperature-measuring hardware can see how the program works, the PDL (0) and PDL (1) statements that obtain the temperature value actually read the game paddles and should be replaced with appropriate PEEK statements in a real biofeedback program.

```
10  GR: COLOR = 9
20  VLIN 37, 39 AT 19: VLIN 37, 39
    AT 21
30  Y = INT(PDL(0)/7)
40  COLOR = 15: VLIN 0, Y AT 20
50  COLOR = 9: VLIN Y+1, 39 AT 20
60  HOME: PRINT 36-Y: GOTO 20
```

This program simulates the operation of a temperature biofeedback apparatus using game paddle 0 in place of a temperature-sensing element. As the game paddle is turned, a video thermometer is displayed in low resolution graphics.

```
10  HGR: HOME: POKE 34, 23: POKE
    35, 24: HOME
20  Y0 = PDL (0): Y1 = PDL (1)
30  IF Y0>159 THEN Y0 = 159
40  IF Y1>159 THEN Y1=159
50  HCOLOR = 0:HPLOT 50,0 TO 50,
    Y0:HPLOT 150,0 TO 150, Y1
60  HCOLOR=3:HPLOT 50, Y0+1 TO
    50,159:HPLOT 150,Y1+1 TO 150,
    159
70  PRINT:HTAB 8:PRINT 160-Y0;
    HTAB 21:PRINT 160-Y1;
80  GOTO 20
```

This program demonstrates how temperature biofeedback could be performed in a competitive setting. High resolution graphics are used to display two video thermometers which rise and fall as game paddles 0 and 1 are turned.

## A Goniometer

The measurement of joint angles is usually performed with a simple device called a goniometer, which is little more than a pair of hinged sticks attached to the center point of a protractor. Measurement of a single joint angle is easy and efficient, but this may not be true for a series of measurements. A computerized goniometer should allow one person to perform a series of joint angle measurements and record the findings at the same time.

The electronics required to make a goniometer is simply a rotary potentiometer connected to an A/D interface. The potentiometer must be attached to the manual goniometer at the hinge so that it is turned as the goniometer is operated. Of course, once the circuit is built it must be calibrated so that the voltage returned by the potentiometer can be converted by software to an angle. Unless the potentiometer accurately returns a voltage that is a mathematical function of its position, the software conversion to degrees of angle should be performed by referring to a table of corresponding values of voltages and angles created during the calibration process.

## A Digitizer

The mechanical analysis of sport performance is a vital component of coaching and athlete development, particularly at the elite level. Usually this is done by projecting successive frames of a high-speed motion picture of a performance on a flat surface and using a digitizer to measure the Cartesian coordinates of the major joints for entry into a computer for analysis. Digitizers are available commercially but can be extremely expensive. A low cost though somewhat slow digitizer can be made using two potentiometers and two channels of an A/D interface to form two goniometers. These are then placed on the projection surface with one arm of each overlapping the other. As Figure 7.6 illustrates, the two overlapping arms and the two free arms of the goniometers form a triangle that is unique for every position on the projection surface. After calibration of each potentiometer, software can be used to calculate the angles and, using trigonometry, the coordinates (Cartesian or polar) of the intersection of the free arms.

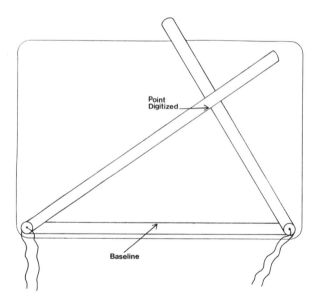

Point Digitized

Baseline

**Figure 7.6** Two potentiometers used to make a digitizer for cinematographic digitization. Initially only the length of the base (b) is known. Angles 1 and 2 are computed through conversion of the voltages passed by the potentiometers. The slopes (m and m') of each side of the triangle are the tangent ratios of angles 1 and 2, respectively. Because the net rise of the two sides with respect to the base must equal zero, the x coordinate of the apex of the triangle is found by solution of the equation

$$m(x) = -m'(b-x)$$

and the y coordinate is then found using the tangent ratio separately for each angle.

The programmer writing a program to perform this solution is warned that the built-in trigonometric functions in many computers expect radians, not degrees, as input.

## EKG, EMG, and EEG Devices

Measurement of the electrical activity produced within the body by muscle and nerve activity is absolutely fundamental to the study of physiology and has in-numerable applications in the sport sciences. Nerve impulses and muscle con-tractions produce electrical activity which can be monitored by measuring the changing potential between two electrodes placed against the skin in different locations. After amplification, these changes can be used to deflect a tracing nee-dle or oscilloscope display or fed to a computer through an A/D interface for further processing.

Because the electrical signals produced by the body are so small, their amplification is not as simple as merely routing them through an Op-Amp cir-cuit. AC signals from simple building wiring and slight imbalances caused by unmatched resistors or sloppy construction techniques would render such an ap-proach unusable or untenable. Furthermore, it is absolutely essential to isolate all current used by the amplifier circuit from the electrodes attached to a human subject, as serious injury or death could result from even small current leakages. A successful and safe circuit for nonclinical demonstration purposes can be built using an instrumentation amplifier and an isolation amplifier circuit. However, because of the complexities and the safety factors involved, no example circuit is presented here.

In any case, most laboratories involved in the study of exercise physiology will already be equipped with the proper equipment for safely and accurately sens-ing and recording EKG, EMG, and/or EEG signals. By using the output of this equipment, which normally drives a tracing needle or other recording or display apparatus, it is possible to computerize the recording and analysis of these mea-sures. For the protection of the A/D interface and the computer, such a project should not be undertaken without consulting a qualified technician or engineer.

# Appendices

## Appendix A

### Table of Decimal, Hexadecimal, and Binary Numbers

| Dec. | Hex. | Binary | Dec. | Hex. | Binary |
|------|------|--------|------|------|--------|
| 0  | 0 | 0     | 61 | 3d | 111101  |
| 1  | 1 | 1     | 62 | 3e | 111110  |
| 2  | 2 | 10    | 63 | 3f | 111111  |
| 3  | 3 | 11    | 64 | 40 | 1000000 |
| 4  | 4 | 100   | 65 | 41 | 1000001 |
| 5  | 5 | 101   | 66 | 42 | 1000010 |
| 6  | 6 | 110   | 67 | 43 | 1000011 |
| 7  | 7 | 111   | 68 | 44 | 1000100 |
| 8  | 8 | 1000  | 69 | 45 | 1000101 |
| 9  | 9 | 1001  | 70 | 46 | 1000110 |
| 10 | A | 1010  | 71 | 47 | 1000111 |
| 11 | B | 1011  | 72 | 48 | 1001000 |
| 12 | C | 1100  | 73 | 49 | 1001001 |
| 13 | D | 1101  | 74 | 4A | 1001010 |
| 14 | E | 1110  | 75 | 4B | 1001011 |
| 15 | F | 1111  | 76 | 4C | 1001100 |
| 16 | 10 | 10000 | 77 | 4D | 1001101 |
| 17 | 11 | 10001 | 78 | 4E | 1001110 |
| 18 | 12 | 10010 | 79 | 4F | 1001111 |
| 19 | 13 | 10011 | 80 | 50 | 1010000 |
| 20 | 14 | 10100 | 81 | 51 | 1010001 |
| 21 | 15 | 10101 | 82 | 52 | 1010010 |
| 22 | 16 | 10110 | 83 | 53 | 1010011 |

(continued on next page)

## Table of Decimal, Hexadecimal, and Binary Numbers (cont.)

| Dec. | Hex. | Binary | Dec. | Hex. | Binary |
|------|------|--------|------|------|--------|
| 23 | 17 | 10111 | 84 | 54 | 1010100 |
| 24 | 18 | 11000 | 85 | 55 | 1010101 |
| 25 | 19 | 11001 | 86 | 56 | 1010110 |
| 26 | 1A | 11010 | 87 | 57 | 1010111 |
| 27 | 1B | 11011 | 88 | 58 | 1011000 |
| 28 | 1C | 11100 | 89 | 59 | 1011001 |
| 29 | 1D | 11101 | 90 | 5A | 1011010 |
| 30 | 1E | 11110 | 91 | 5B | 1011011 |
| 31 | 1F | 11111 | 92 | 5C | 1011100 |
| 32 | 20 | 100000 | 93 | 5D | 1011101 |
| 33 | 21 | 100001 | 94 | 5E | 1011110 |
| 34 | 22 | 100010 | 95 | 5F | 1011111 |
| 35 | 23 | 100011 | 96 | 60 | 1100000 |
| 36 | 24 | 100100 | 97 | 61 | 1100001 |
| 37 | 25 | 100101 | 98 | 62 | 1100010 |
| 38 | 26 | 100110 | 99 | 63 | 1100011 |
| 39 | 27 | 100111 | 100 | 64 | 1100100 |
| 40 | 28 | 101000 | 101 | 65 | 1100101 |
| 41 | 29 | 101001 | 102 | 66 | 1100110 |
| 42 | 2A | 101010 | 103 | 67 | 1100111 |
| 43 | 2B | 101011 | 104 | 68 | 1101000 |
| 44 | 2C | 101100 | 105 | 69 | 1101001 |
| 45 | 2D | 101101 | 106 | 6A | 1101010 |
| 46 | 2E | 101110 | 107 | 6B | 1101011 |
| 47 | 2F | 101111 | 108 | 6C | 1101100 |
| 48 | 30 | 110000 | 109 | 6D | 1101101 |
| 49 | 31 | 110001 | 110 | 6E | 1101110 |
| 50 | 32 | 110010 | 111 | 6F | 1101111 |
| 51 | 33 | 110011 | 112 | 70 | 1110000 |
| 52 | 34 | 110100 | 113 | 71 | 1110001 |
| 53 | 35 | 110101 | 114 | 72 | 1110010 |
| 54 | 36 | 110110 | 115 | 73 | 1110011 |
| 55 | 37 | 110111 | 116 | 74 | 1110100 |
| 56 | 38 | 111000 | 117 | 75 | 1110101 |
| 57 | 39 | 111001 | 118 | 76 | 1110110 |
| 58 | 3A | 111010 | 119 | 77 | 1110111 |
| 59 | 3B | 111011 | 120 | 78 | 1111000 |
| 60 | 3C | 111100 | 121 | 79 | 1111001 |
| 122 | 7A | 1111010 | 181 | B5 | 10110101 |
| 123 | 7B | 1111011 | 182 | B6 | 10110110 |
| 124 | 7C | 1111100 | 183 | B7 | 10110111 |

(continued on next page)

## Table of Decimal, Hexadecimal, and Binary Numbers (cont.)

| Dec. | Hex. | Binary | Dec. | Hex. | Binary |
|------|------|--------|------|------|--------|
| 125 | 7D | 1111101 | 184 | B8 | 10111000 |
| 126 | 7E | 1111110 | 185 | B9 | 10111001 |
| 127 | 7F | 1111111 | 186 | BA | 10111010 |
| 128 | 80 | 10000000 | 187 | BB | 10111011 |
| 129 | 81 | 10000001 | 188 | BC | 10111100 |
| 130 | 82 | 10000010 | 189 | BD | 10111101 |
| 131 | 83 | 10000011 | 190 | BE | 10111110 |
| 132 | 84 | 10000100 | 191 | BF | 10111111 |
| 133 | 85 | 10000101 | 192 | C0 | 11000000 |
| 134 | 86 | 10000110 | 193 | C1 | 11000001 |
| 135 | 87 | 10000111 | 194 | C2 | 11000010 |
| 136 | 88 | 10001000 | 195 | C3 | 11000011 |
| 137 | 89 | 10001001 | 196 | C4 | 11000100 |
| 138 | 8A | 10001010 | 197 | C5 | 11000101 |
| 139 | 8B | 10001011 | 198 | C6 | 11000110 |
| 140 | 8C | 10001100 | 199 | C7 | 11000111 |
| 141 | 8D | 10001101 | 200 | C8 | 11001000 |
| 142 | 8E | 10001110 | 201 | C9 | 11001001 |
| 143 | 8F | 10001111 | 202 | CA | 11001010 |
| 144 | 90 | 10010000 | 203 | CB | 11001011 |
| 145 | 91 | 10010001 | 204 | CC | 11001100 |
| 146 | 92 | 10010010 | 205 | CD | 11001101 |
| 147 | 93 | 10010011 | 206 | CE | 11001110 |
| 148 | 94 | 10010100 | 207 | CF | 11001111 |
| 149 | 95 | 10010101 | 208 | D0 | 11010000 |
| 150 | 96 | 10010110 | 209 | D1 | 11010001 |
| 151 | 97 | 10010111 | 210 | D2 | 11010010 |
| 152 | 98 | 10011000 | 211 | D3 | 11010011 |
| 153 | 99 | 10011001 | 212 | D4 | 11010100 |
| 154 | 9A | 10011010 | 213 | D5 | 11010101 |
| 155 | 9B | 10011011 | 214 | D6 | 11010110 |
| 156 | 9C | 10011100 | 215 | D7 | 11010111 |
| 157 | 9D | 10011101 | 216 | D8 | 11011000 |
| 158 | 9E | 10011110 | 217 | D9 | 11011001 |
| 159 | 9F | 10011111 | 218 | DA | 11011010 |
| 160 | A0 | 10100000 | 219 | DB | 11011011 |
| 161 | A1 | 10100001 | 220 | DC | 11011100 |
| 162 | A2 | 10100010 | 221 | DD | 11011101 |
| 163 | A3 | 10100011 | 222 | DE | 11011110 |
| 164 | A4 | 10100100 | 223 | DF | 11011111 |
| 165 | A5 | 10100101 | 224 | E0 | 11100000 |

(continued on next page)

## Table of Decimal, Hexadecimal, and Binary Numbers (cont.)

| Dec. | Hex. | Binary | Dec. | Hex. | Binary |
|------|------|----------|------|------|----------|
| 166 | A6 | 10100110 | 225 | E1 | 11100001 |
| 167 | A7 | 10100111 | 226 | E2 | 11100010 |
| 168 | A8 | 10101000 | 227 | E3 | 11100011 |
| 169 | A9 | 10101001 | 228 | E4 | 11100100 |
| 170 | AA | 10101010 | 229 | E5 | 11100101 |
| 171 | AB | 10101011 | 230 | E6 | 11100110 |
| 172 | AC | 10101100 | 231 | E7 | 11100111 |
| 173 | AD | 10101101 | 232 | E8 | 11101000 |
| 174 | AE | 10101110 | 233 | E9 | 11101001 |
| 175 | AF | 10101111 | 234 | EA | 11101010 |
| 176 | B0 | 10110000 | 235 | EB | 11101011 |
| 177 | B1 | 10110001 | 236 | EC | 11101100 |
| 178 | B2 | 10110010 | 237 | ED | 11101101 |
| 179 | B3 | 10110011 | 238 | EE | 11101110 |
| 180 | B4 | 10110100 | 239 | EF | 11101111 |
| 240 | F0 | 11110000 | | | |
| 241 | F1 | 11110001 | | | |
| 242 | F2 | 11110010 | | | |
| 243 | F3 | 11110011 | | | |
| 244 | F4 | 11110100 | | | |
| 245 | F5 | 11110101 | | | |
| 246 | F6 | 11110110 | | | |
| 247 | F7 | 11110111 | | | |
| 248 | F8 | 11111000 | | | |
| 249 | F9 | 11111001 | | | |
| 250 | FA | 11111010 | | | |
| 251 | FB | 11111011 | | | |
| 252 | FC | 11111100 | | | |
| 253 | FD | 11111101 | | | |
| 254 | FE | 11111110 | | | |
| 255 | FF | 11111111 | | | |

# Appendix B

## Selected Digital Integrated Circuits

| Number* | Function | No. of Pins |
|---|---|---|
| 4001 | Quad 2-Input NOR Gate | 14 |
| 4009 | Hex Inverter | 16 |
| 4012 | Dual 4-Input NAND Gate | 14 |
| 4013 | Dual D Type Flip-Flop | 14 |
| 4017 | Decoded Decade Counter | 16 |
| 4022 | Decoded Octal Counter | 16 |
| 4029 | Presettable 4-Bit/Decimal Counter | 16 |
| 4070 | Quad 2-Input EOR Gate | 14 |
| 4093 | Quad 2-Input NAND Schmitt Trigger | 14 |
| 4099 | 8-Bit Addressable Latch | 16 |
| 40106 | Hex Schmitt Trigger | 14 |
| 40192 | 4 Bit Up/Down Counter | 16 |
| 4508 | Dual 4-Bit Latch | 24 |
| 7400 | Quad 2-Input NAND Gate | 14 |
| 7402 | Quad 2-Input NOR Gate | 14 |
| 7404 | Hex Inverter | 14 |
| 7408 | Quad 2-Input AND Gate | 14 |
| 7413 | Dual 4-Input NAND Schmitt Trigger | 14 |
| 7414 | Hex Schmitt Trigger | 14 |
| 7420 | Dual 4-Input NAND Gate | 14 |
| 7422 | Dual 4-Input NAND Gate | 14 |
| 7432 | Quad 2-Input OR Gate | 14 |
| 7447 | 7 Segment LED Decoder/Driver | 16 |
| 7448 | 7 Segment LED Decoder/Driver | 16 |
| 7486 | Quad 2-Input EOR Gate | 14 |
| 7490 | Decade Counter | 14 |
| 7493 | Hexidecimal Counter | 14 |
| 74121 | One Shot | 14 |
| 74122 | Retriggerable One Shot | 14 |
| 74123 | Dual Retriggerable One Shot | 16 |
| 74132 | Quad 2-Input NAND Schmitt Trigger | 14 |
| 74138 | 3 to 8 Line Decoder | 16 |
| 74154 | 4 to 16 i Line Decoder | 24 |
| 74177 | Presettable Binary Counter | 14 |
| 74196 | Presettable Binary Counter | 14 |
| 8301 | 1 of 10 Decoder | 16 |
| 8311 | 4 to 16 Line Decoder | 24 |

*Frequently the same device is available using different technology and this is designated by letters within the logic number. The 7400, 74LS00, and 74C00, for example, all have identical functions and pinouts but are manufactured using TTL, LSTTL, and CMOS technology.

# Appendix C

## Resistor Color Codes

Bands 1 and 2 are Digits

0 = Black
1 = Brown
2 = Red
3 = Orange
4 = Yellow
5 = Green
6 = Blue
7 = Violet
8 = Gray
9 = White

Band 3 is Multiplier

.01 = Silver
.1 = Gold
1 = Black
10 = Brown
100 = Red
1000 = Orange
10000 = Yellow
100000 = Green
1000000 = Blue

Last Band is Tolerance

5% = Gold
10% = Silver
20% = No Band

# Index